MW00416113

ACS SYMPOSIUM SERIES 401

Nucleotide Analogues as Antiviral Agents

John C. Martin, EDITOR
Bristol-Myers Company

Developed from a symposium sponsored
by the Division of Carbohydrate Chemistry
and the Division of Medicinal Chemistry
at the 196th National Meeting
of the American Chemical Society,
Los Angeles, California,
September 25-30, 1988

American Chemical Society, Washington, DC 1989

Library of Congress Cataloging-in-Publication Data

Nucleotide analogues as antiviral agents
John C. Martin, editor

Developed from a symposium sponsored by the Divisions of Carbohydrate Chemistry and Medicinal Chemisty at the 196th National Meeting of the American Chemical Society, Los Angeles, California, September 25–30. 1988.

p. cm.—(ACS Symposium Series, 0097–6156; 401).
Includes bibilographies and index.

ISBN 0–8412–1659–2
1. Antiviral agents—Testing—Congresses.
2. Nucleotides—Derivatives—Therapeutic use—Testing—Congresses.

I. Martin, John C., 1951– . II. American Chemical Society. Division of Carbohydrate Chemistry.
III. American Chemical Society. Division of Medicinal Chemistry. IV. American Chemical Society. Meeting (196th: 1988: Los Angeles, Calif.). V. Series

RM411.N83 1989
616.9'25061—dc20 89–15114
CIP

PRINTED IN THE UNITED STATES OF AMERICA

ACS Symposium Series

M. Joan Comstock, *Series Editor*

Foreword

The ACS SYMPOSIUM SERIES was founded in 1974 to provide a medium for publishing symposia quickly in book form. The format of the Series parallels that of the continuing ADVANCES IN CHEMISTRY SERIES except that, in order to save time, the papers are not typeset but are reproduced as they are submitted by the authors in camera-ready form. Papers are reviewed under the supervision of the Editors with the assistance of the Series Advisory Board and are selected to maintain the integrity of the symposia; however, verbatim reproductions of previously published papers are not accepted. Both reviews and reports of research are acceptable, because symposia may embrace both types of presentation.

Contents

Preface

LARGELY IN RESPONSE TO THE AIDS EPIDEMIC, the amount of research directed toward the discovery of antiviral agents has grown dramatically over the last several years. Nucleoside analogues, such as acyclovir and zidovudine, are the most common approved drugs for the treatment of viral infections. In general, these substances have a mechanism of action that involves first phosphorylation in cells to nucleotide analogues and then inhibition of an essential viral enzyme, DNA polymerase. Analogues of nucleotides (phosphorylated derivatives of nucleosides) are often negatively charged. In the past they were thought to have little potential for efficacy because of the low permeability of cells to charged molecules. Contrary to this expectation, several negatively charged nucleotide analogues have been found during the last five years to exert potent in vivo antiviral effects.

This new awareness of the potential therapeutic utility of nucleotide analogues has led a number of laboratories to develop research programs on this topic. The contributors to *Nucleotide Analogues as Antiviral Agents* are an international group of scientists carrying out research at the forefront of nucleotide drug design. The individual chapters describe many new concepts and results, much of the information unpublished to date. The goal of this book is to provide any chemist, biochemist, or biologist who is interested or involved in antiviral research with a collection of information that will offer insight into this increasingly important research topic.

The first chapter provides an overview of the field and also considerable new structure–activity data on analogues of pyrophosphate. Chapters 2–6 describe the chemistry and biological activities of a number of stable acyclic nucleotide analogues that have in vivo activity against herpesviruses and retroviruses. Progress in biochemical and molecular biological research has advanced our understanding of virus replication and resulted in the identification of new targets for inhibition, in addition to the viral DNA polymerases. Chapters 7–9 describe nucleotide analogues that inhibit viral thymidine kinase, glycosylation of viral proteins, and aminoacyl-tRNA synthetase, respectively. Chapter 10 provides, by example, means to design nucleoside analogues to be phosphorylated to nucleotides by host enzymes, or in some cases, analogues that are resistant to this phosphorylation. Chapter 11 details the evaluation of nucleotide dimers of nucleosides active against human immunodeficiency virus. The final chapter of this book contains a discussion of the current efforts to design stable oligonucleotides as antisense inhibitors of the transcription or translation of key viral genes.

To date, no nucleotide analogues have been approved for use as antiviral drugs. However, a number of the substances described in this book have the potential to be developed as antiviral therapies. As our knowledge of the biological properties (potency, mechanism of action, toxicity, pharmacokinetics, and metabolism) of nucleotide analogues increases, the rational design of superior antiviral agents should become increasingly successful.

Acknowledgments

I am very appreciative of the excellent and timely contributions made by authors of chapters in this book. I am also grateful to the many experts in the field of antiviral research who contributed their time to the evaluation of the individual chapters in order to make valuable suggestions for improvements and to Cheryl Shanks of the ACS Books Department. David C. Baker of the Division of Carbohydrate Chemistry and William T. Comer of the Division of Medicinal Chemistry of the American Chemical Society were instrumental in arranging for the financial support of their respective divisions. Also, David C. Baker was especially helpful in his support and advice concerning the organization of the symposium and this book.

JOHN C. MARTIN
Bristol-Myers Company
Wallingford, CT 06492–7660

May 18, 1989

Chapter 1

Inhibitors of Viral Nucleic Acid Polymerases

Pyrophosphate Analogues

Charles E. McKenna[1], Jeffrey N. Levy[1], Leslie A. Khawli[1,3], Vahak Harutunian[2], Ting-Gao Ye[1,4], Milbrey C. Starnes[2], Ashok Bapat[2,5], and Yung-Chi Cheng[2,6]

[1]Department of Chemistry, University of Southern California, Los Angeles, CA 90089–0744

[2]Department of Pharmacology and Medicine, School of Medicine, University of North Carolina–Chapel Hill, Chapel Hill, NC 27514

Virus-specific enzymes essential for viral nucleic acid replication or related functions are targets for inhibition by substrate or product nucleotide analogues in which one or more P-O bonds are replaced by a P-C bond. The simplest examples of these are PFA (phosphonoformic acid) and PAA (phosphonoacetic acid), representing analogues of 'pyrophosphate' moieties in nucleotides. The synthesis of a series of α-halogenated and α-oxo PAA and MDP (methanediphosphonate) derivatives is described and structure/activity relationships in their inhibition of several human (α,β,γ) and viral (HSV, EBV, HIV) DNA polymerases are presented. Inhibition of HIV RNA-directed DNA polymerase (reverse transcriptase) by PFA, α-oxophosphonoacetate and α-oxomethanediphosphonate is shown to be pH- and template-dependent. Combination of phosphonoacetate derivatives and anti-viral nucleosides into 'hybrid' nucleotide analogues is briefly discussed.

Viruses, infecting and reproducing within host cells, have long been an elusive target for chemotherapy. However, recent advances in

[3]Current address: Department of Pathology, School of Medicine, University of Southern California, Los Angeles, CA 90033

[4]Current address:Wenzhou Institute of Pesticide Research, Huiqiaopu, Wenzhou, Zhejiang, China

[5]Current address: Wyeth Laboratories, P.O. Box 8299, Philadelphia, PA 19101–8299

[6]Current address: Department of Pharmacology, School of Medicine, Yale University, New Haven, CT 06510

0097–6156/89/0401–0001$06.00/0

molecular virology have increased optimism about the feasibility of creating rationally designed, effective and non-toxic anti-viral agents (1). Virus-encoded gene products required for DNA replication can significantly differ in substrate specificity from normal DNA polymerases involved in host cell reproduction. For example, *Herpes simplex* viruses 1 and 2 (HSV-1, HSV-2) induce synthesis of DNA polymerases having Km values for deoxynucleoside 5'-triphosphate substrates (dNTP's) that are smaller ($\sim 10^{-7}$M) than the Km values of any known mammalian DNA polymerase (2). This confers an advantage to the virus in competing for intracellular substrates, but also renders it vulnerable to selective inhibition. A similar rationale underlies ongoing efforts to develop inhibitors for other types of virus-specific nucleic acids polymerases, such as the RNA polymerase of influenza viruses and the RNA-dependent DNA polymerase (reverse transcriptase) of Human Immunodeficiency Virus (HIV). Virus-specific enzymes involved in related functions, such as nucleoside kinases and integrases, are also potential drug targets.

Virus-specific DNA polymerases catalyze DNA- or RNA-specified condensation of a nucleoside 5'-triphosphate with a growing polynucleotide strand, eliminating pyrophosphate (Scheme 1). dNTP substrates, and also the pyrophosphate byproduct, are recognized starting points for inhibitor design. Modification of mononucleotide may focus on the purine or pyrimidine base, the sugar, the triphosphate, or several moieties together. The resulting analogue could be intended to interact reversibly or irreversibly with the targeted viral DNA polymerase. Nucleotides lacking a 3' hydroxyl group can cause chain termination, resulting in product inhibition. In addition to inhibitor potency, properties important for useful anti-viral activity include selectivity relative to similar host enzymes, and ability to penetrate cell membranes.

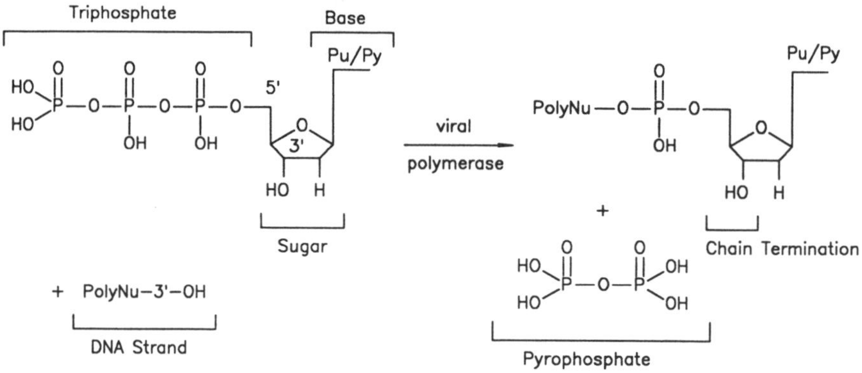

Triphosphate
Base
Pu/Py
5'
3'
HO H
Sugar
+ PolyNu–3'–OH
DNA Strand
viral polymerase
PolyNu—O—P—O—
Pu/Py
HO H
Chain Termination
+
Pyrophosphate

SCHEME 1

Pyrophosphate analogues might also be thought of as fragmentary nucleotides, with only the oligophosphate moiety mimicked. Perhaps the structurally simplest compound discovered to have significant anti-viral activity, phosphonoformic acid (PFA, 1), belongs to this category. PFA is believed to interact directly with viral polymerases, interfering with substrate binding and thus blocking replication of DNA. Phosphonate analogues of pyrophosphate, which contain P-C bonds in place of P-O bonds, will be introduced in more detail following this background section.

More recently discovered *nucleoside* anti-viral agents such as Acyclovir (ACV, 9-(2-hydroxyethoxymethyl)guanine; active against HSV) and AZT (3'-azido-2',3'-dideoxythymidine; active against HIV) require conversion to nucleoside triphosphates by viral and/or cellular kinases for activity. These drugs are thus liposoluble precursors of corresponding nucleotide analogues and are believed to inhibit the viral polymerases by competing with natural dNTP substrates for a common binding site. ACV and AZT triphosphate also exert anti-viral activity by functioning as viral DNA polymerase mononucleotide substrates, leading to their incorporation into the bound product DNA strand where they act as chain terminators. This dual inhibition mechanism is outlined for ACV in Scheme 2, where ACV-MP and ACV-TP are ACV mono- and triphosphate, respectively, HSV-DP represents a *Herpes simplex* DNA polymerase, and DNA-ACV-MP is the product DNA strand terminated by ACV-MP, which has no 3' hydroxyl group. The employment of a nucleoside prodrug circumvents two disadvantages inherent in direct use of the corresponding nucleoside triphosphate: its anionic charge, which limits cell transport; and the susceptibility of its triphosphate group to enzymatic hydrolysis. An activation process solely dependent on host cellular enzymes is indicated for AZT (3), but a viral thymidine kinase (TK) has been implicated in the initial phosphorylation of ACV to ACV-MP (4,5) Obligatory activation by another viral enzyme amplifies the selectivity of the inhibitor (and presents an additional anti-viral drug target (6)). However, nucleoside analogues phosphorylated selectively by viral TK on the path to their active triphosphate forms may have a restricted spectrum of activity due to virus-dependent variation in TK specificity (7). ACV-resistant HSV isolates often have a TK^- phenotype, whereas mutants with alterations in DNA polymerase appear to arise with lower frequency (8).

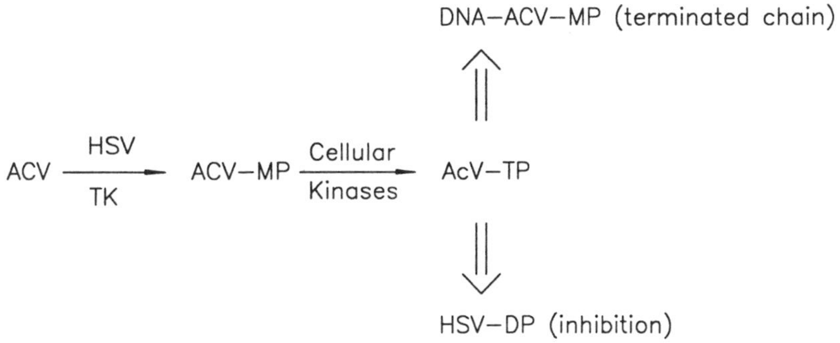

DNA–ACV–MP (terminated chain)

⇑

ACV —(HSV TK)→ ACV–MP —(Cellular Kinases)→ AcV–TP

⇓

HSV–DP (inhibition)

SCHEME 2

Nucleotide anti-viral analogues would be presumed to affect the target polymerase directly, bypassing activation or, in the case of nucleoside monophosphate analogues, requiring only cellular kinases for further phosphorylation (9). Virus-infected cells are often more permeable to nucleotides than normal cells, which might mitigate the potential problem of limited cell transport due to ionic charge, and nucleotide analogues in which labile P-O bonds are replaced by P-C or other bonds resistant to enzymatic hydrolysis may be less toxic and teratogenic than corresponding nucleosides (10). In support of this idea, a phosphonate analogue of DHPG (9-[(1,3-dihydroxy-2-propoxy)methyl]guanine) monophosphate was found to exhibit substantial activity against Cytomegalovirus (CMV) with significantly lower toxicity than corresponding mono- and diphosphate DHPG inhibitors (11). Active interest in nucleotides as anti-viral agents is relatively recent, although other uses of phosphonates as biophosphate analogues have been known for some time (12).

Oligonucleotide analogues constitute a fourth category of nucleotide anti-viral agent, designed e.g. to block viral gene expression at the level of transcription or translation. These sequence-specific nucleic acid segments are typically intended to hybridize with complementary viral template ("anti-sense" inhibition), and generally have modified 3′,5′ phosphate linkages to improve stability to nucleases and enhance transport (13).

The viral inhibitor design strategies summarized above are reflected in the different contributions comprising this volume. Our paper will focus primarily on pyrophosphate analogues as such, and (briefly) as components of 'hybrid' nucleotide analogues.

Pyrophosphate Analogues

The earliest pyrophosphate analogue found to possess anti-viral activity was phosphonoacetic acid (PAA, **2a**) (14) (for convenience, structural references for the analogues of this type are given in fully protonated forms (Scheme 3); the actual inhibitors are assumed to be corresponding anionic forms). Both PFA and PAA inhibit replication of HSV-1 and HSV-2 and suppress initial herpes lesions when applied topically (14,15). PFA is more potent than PAA against HSV although IC_{50} values against HSV-induced DNA polymerases are similar for the two compounds, suggesting that multiple factors are involved in overall drug effectiveness.

Modifications in the phosphonate/carboxylate groups of PAA and PFA by esterification with simple alkyl or aryl groups or by replacement with other combinations of acidic functional groups have generally resulted in less active compounds (15), although exceptions have been reported (16,17). Prior studies of PAA-like compounds (15-20) provide evidence that one phosphonate group, in close proximity (preferably one or zero intervening atoms) to a carboxylate, is associated with activity. Replacement of the carboxylate group in **2a** by a phosphonate group (methanediphosphonic acid/methylene bis[phosphonic acid], MDP, **3a**) resulted in loss of activity (18). From the perspective of providing access to derivatives useful for probing structure-function relationships, the methylene group present in PAA offers substitutional latitude not available in PFA, but α-substitution of PAA has usually decreased activity (15,17-18). Apart from this apparent steric effect, and the terminal-group correlations reviewed above, structure/activity relationships for pyrophosphate analogues are not well understood.

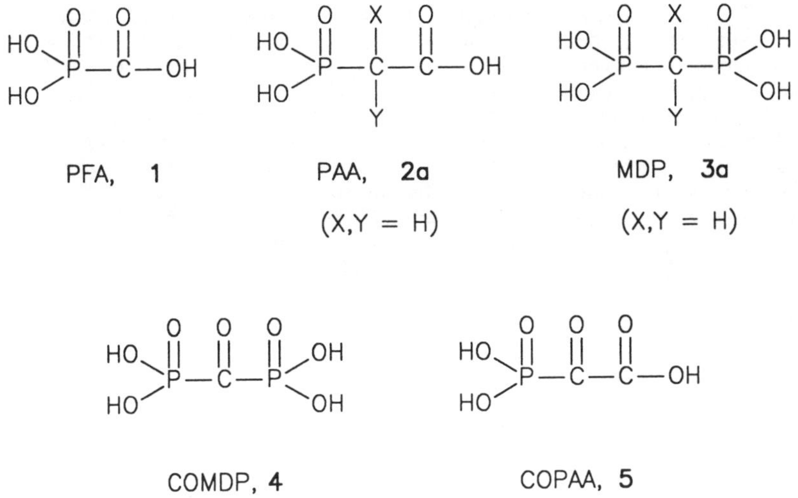

PFA, **1** PAA, **2a** (X,Y = H) MDP, **3a** (X,Y = H)

COMDP, **4** COPAA, **5**

SCHEME 3

α-Halogenated Pyrophosphate Analogues. Fluorophosphonoacetic and difluorophosphonoacetic acids (**2b**, **2c**) were previously discussed as possible new viral inhibitors (21). The polar -CHF- and $-CF_2-$ groups in **2b** and **2c** more closely mimic the anhydride oxygen in P_i-O-P_i than does the $-CH_2-$ group of PAA, at a cost of relatively small steric perturbation, but the net effect of such modification was not clearly predictable. An analytical advantage of α-fluoromethylene analogues is the presence of the ^{19}F nucleus as a potentially useful NMR probe.

To establish systematically the effect of halogen α-substitution on the activity of PAA towards particular viral polymerases, we prepared an integral set of α-halo analogues: XYPAA, X,Y = H,F (**2b**); F,F (**2c**); H,Cl (**2d**); Cl,Cl (**2e**); H,Br (**2f**); Br,Br (**2g**); F,Cl (**2h**); F,Br (**2i**); Cl,Br (**2j**); CH_3,F (**2l**); CH_3,Cl (**2m**); CH_3,Br (**2n**). Reported inhibition of an RNA virus polymerase by ClMDP (**3d**), Cl_2MDP (**3e**) and Br_2MDP (**3g**), but not by MDP itself (22,23) prompted us to include in our inhibition studies a comparable set of α-halo methanediphosphonates (XYMDP, **3b**-**3j**).

α-Oxo Pyrophosphate Analogues. PAA, in contrast to its ability to inhibit viral DNA polymerases, was recently found to be ineffective as an inhibitor of HIV-1 reverse transcriptase, whereas PFA was a potent inhibitor of this enzyme (24). It was also shown that α-oxomethanediphosphonate (carbonyldiphosphonate, COMDP, **4**) is a moderately good inhibitor of reverse transcriptase, whereas the corresponding methylene compound **3a** is inactive (24). In these studies artificial homonucleotide templates were used in reverse transcriptase inhibition assays, and considerable variation in the inhibitory activity of individual compounds was observed with different templates. The existence of a ketone-hydrate equilibrium for **4** in aqueous solution, potentially complicating identification of the actual inhibitory species, was not addressed.

Commonality of an α-carbonyl group in the reverse transcriptase inhibitors **1** and **4** led us to synthesize and determine the reverse transcriptase inhibition activity of the α-keto analogue of PAA, α-oxophosphonoacetic acid (phosphonoglyoxalic acid, COPAA, **5**). The influence of assay template on observed inhibitor activity was examined for **1** and **4**, and the effect of pH on reverse transcriptase inhibition by **1**, **4** and **5** was also investigated.

Synthetic Aspects.

α-Halo Phosphonates. The 9 triethyl esters (**6b-j**) were prepared from a single precursor, triethylphosphonoacetate (**6a**) (21,25; McKenna, C.E. et al., *J. Fluorine Chem.*, in press) (Scheme 4). The dichloro and dibromo esters (**6e**, **6g**) were made by hypohalogenation ($NaOCl$ or $NaOH/Br_2$) of **6a**, and difluoro ester **6c** was obtained with the monofluoro product **6b** by treatment of **6a** with tBuOK followed by $FClO_3$. Reduction of **6e** (Na_2SO_3) and **6g** (Sn^{2+}) provided the corresponding monochloro and monobromo esters **6d** and **6f**.

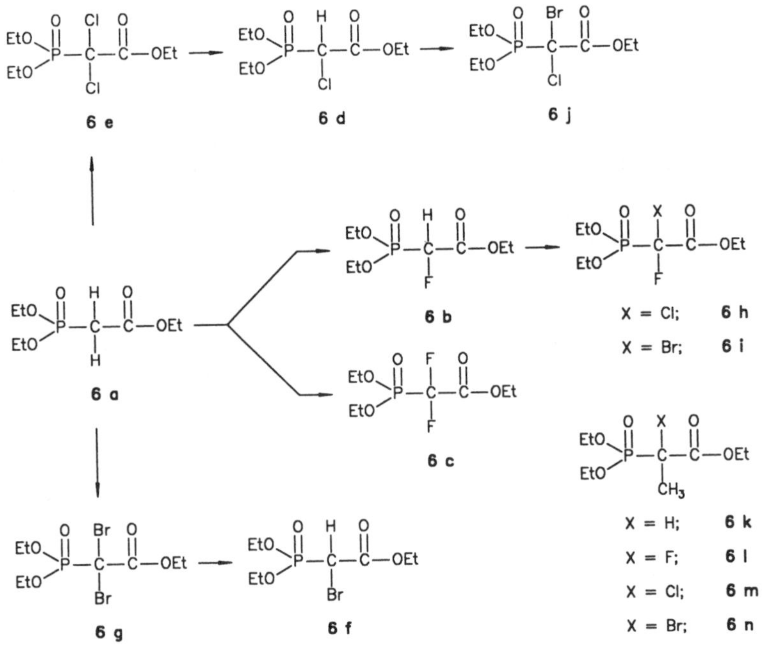

SCHEME 4

Similar hypohalogenation procedures yielded the mixed dihalo esters **6h** and **6i** from **6b**, and **6j** from **6d**. α-Methyl α-halo esters **6l**-**6n** were synthesized by analogous methods from triethyl 2-phosphonopropionate **6k**.

The corresponding acids **2b**-**2j** and **2l**-**2n** were prepared by refluxing the esters in conc. HCl for 6 h and were isolated as dicyclohexylamine (DCHA) or pyridine (Pyr) salts (21; McKenna, C.E. et al., *J. Fluorine Chem.*, in press; (26). We find that an improved yield of BrPAA (**2f**) free from traces of ClPAA can be obtained by replacing the HCl by aqueous HBr (25%, reflux, 45 min); also, yields of dihalo acids **2g** and **2j** can be optimized by adjusting the reflux time in HCl (25% recommended) to the minimum necessary (ca. 1/2 h) (Ye, T.-G., unpublished).

Investigation of tetramethyl, tetraethyl and tetraisopropyl MDP as a common synthetic origin for all nine XYMDP derivatives revealed that tetraisopropyl MDP is preferable for this purpose. Synthetic routes similar to those outlined above for **6b**-**6j** were used to prepare the set of tetraisopropyl esters corresponding to α-halo methanediphosphonic acids **3b**-**3j**, which were then obtained by refluxing the appropriate ester in HCl (27).

α-Keto Phosphonates.

Our synthetic studies of **5** and its triethyl ester **7** were recently communicated (28). A purported preparation of ester **7** *via* Michaelis-Arbuzov reaction between ethyl oxalyl chloride and triethyl phosphite (29) in our hands led to other final products. Attempted alkaline hydrolysis of Cl_2PAA (**2e**) to **5**, by analogy to the preparation of carbonyldiphosphonate **4** from **3e** (30), led to decomposition; methods used to oxidize diethyl malonate to diethyl oxomalonate (31) could not be readily extended to **6a**. Alternative oxidative pathways starting from triethyl phosphonoacrylate using ozonolysis, RuO_4/IO_4^- and RuO_4/ClO^- were also explored (Levy, J.N.; McKenna, C.E., *Phosphorus Sulfur*, in press). Our preferred route to **5** exploits carbene-mediated oxygen transfer chemistry originally developed for deprotection of alkenes protected as epoxides (32): thermal decomposition of triethyl diazophosphonoacetate (33) **8** catalyzed by rhodium (II) acetate in the presence of propylene oxide gives **7** in good yield.

However, we were unable to convert **7** to **5** by direct hydrolysis with HCl or HBr, owing to the reactivity of its α-carbonyl group, which also quantitatively adds H_2O. Silyldealkylation using chloro-, bromo- or iodotrimethylsilane also proved unsatisfactory. We circumvented this problem using a synthetic sequence in which P-OR silyldealkylation precedes the oxytranfer step, followed by self-catalyzed acid hydrolysis of the remaining (carboxylate) ester group, and isolation of the ketone **5** as an amine salt (Scheme 5). Thus, the free acid was prepared by silyldealkylation of **8** with bromotrimethylsilane (34) to ethyl P,P-bis(trimethylsilyl) diazophosphonoacetate **9**, oxygenation as described above to the oxophosphonoacetate mixed ester **10** and treatment with water at 25 °C to selectively hydrolyze the trimethylsilyl groups (**11** and its hydrate). Heating the resulting solution to 56 °C for 26 h removed the carboxyl ethyl group, giving the ketone **5** in equilibrium with its hydrate **12**. The phosphonic monoacid of **5** was recovered as the bis(dicyclohexylammonium) salt.

Sodium carbonyldiphosphonate **4** was prepared by a slight modification of a published method (30).

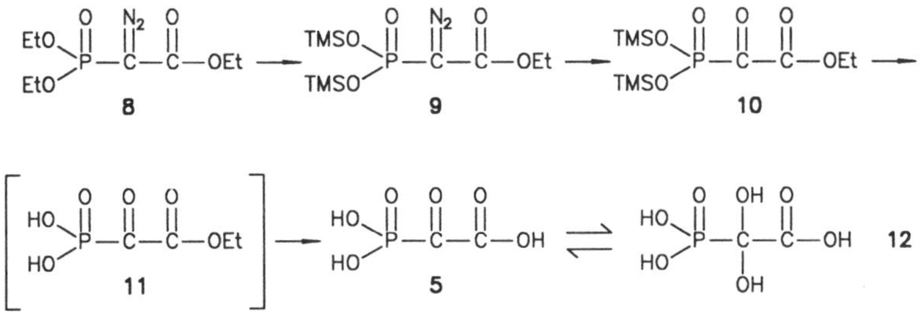

SCHEME 5

Ketone-Hydrate Equilibria of 4 and 5. Unlike the triester 7, 5 has an equilibrium with its hydrate which is *pH-dependent* (28; McKenna, C.E.; Levy, J.N., in preparation). In the trianion, negative charge stabilizes the ketone form, which predominates (> 99%) in H_2O. Protonations successively lower the charge, activating the α-carbonyl and thus producing more hydrate. At pH 7-8, the α-carbonyl reactivity is higher (significant fraction of ketone present as dianion), but not sufficiently to favor the hydrate (present as a few percent at room temperature). Below pH 6, the hydrate becomes the major species. α-Oxomethanediphosphonate 4 displays similar behavior but with relatively greater preference for the keto form at a given pH.

Biochemical Aspects.

Preparation of DNA Polymerases. Activated calf thymus DNA, DNA polymerases from HSV-1 and HSV-2, EBV (Epstein-Barr virus), and peripheral blasts from chronic lymphocytic leucophoresed patients undergoing blast crisis were prepared by previously published methods (35-38). Generally, DNA polymerases were purified by sequential chromatography on DEAE-cellulose, phosphocellulose, and single- or double-stranded-DNA cellulose. The purified enzymes were dialyzed against and stored in 50 mM Tris-HCl (pH 7.5) containing 1 mM each of DTT, EDTA and PMSF, plus 30% glycerol. Purification of reverse transcriptase from HIV-1 will be described elsewhere (Starnes, M.C.; Cheng, Y.-C., *J. Biol. Chem.*, in press).

Herpesvirus and Human DNA Polymerase Assays. Standard viral HSV and EBV polymerase reaction mixtures contained the following: 50 mM Tris-HCl, pH 8.0; 4 mM $MgCl_2$; 0.5 mM dithiothreitol (DTT); 0.2 mg/ml bovine serum albumin; 0.15 M KCl; 0.25 mg/ml activated calf thymus DNA; 0.1 mM each of dATP, dCTP and dGTP; and 10 μM [^{3}H]TTP in a final reaction volume of 50 μl. The reaction was started by adding the enzyme to the reaction mixture and allowed to proceed for 20 min at 37°C. Samples were spotted onto 2.1 cm GF/A filter discs and processed to determine trichloroacetic acid insoluble, filter-bound radioactivity. When determining the inhibitory action of PAA analogs, the reaction mixture containing the appropriate amount of analogue was kept on ice before initiation. Assays with human DNA polymerase α were done similarly except that the pH of the reaction mixture was 7.5 and contained no KCl. For β and γ DNA polymerases the reaction mixture included 100 mM KCl.

HIV-1 Reverse Transcriptase Assays. Standard assays were run at 37°C and contained: 50 mM Tris, pH 8.0, 0.5 mM DTT, 8 mM $MgCl_2$, 100 μg/mL BSA, 150 μg/mL gapped calf thymus DNA, 100 μM each dATP, dCTP, dGTP, 10 μM $[^3H]$-dTTP, and 1-5 μL enzyme in a final volume of 50 μL. Modified assays for pH dependence inhibition studies with PFA (**1**) and α-oxo phosphonates (**4** and **5**) contained 50 mM Hepes, pH 8.2 - 6.5, 8 mM $MgCl_2$, 100 mM KCl, 100 μg/ml BSA, 0.5 A_{260} units/ml of poly(rA)·$(dT)_{10}$, 100 μM $[^3H]$dTTP, and 1-5 μL enzyme in a final volume of 50 μL. Samples were processed as described above.

Herpesvirus DNA Polymerase Inhibition Studies.

α-Halo Phosphonates. Compounds **2b-2j** and **2l-2n** were evaluated as inhibitors of HSV-1, HSV-2, and EBV DNA polymerases vs. human α, β and γ DNA polymerases. IC_{50} values (inhibitor concentrations giving 50% inhibition), including data for PAA and PFA controls, were previously reported in a preliminary communication (26) and are presented in Table I. Significant inhibition of herpesviral polymerases was observed with FPAA (IC_{50} 2-4 μM (HSV-1, HSV-2); 14 μM (EBV)), ClPAA (IC_{50} 3-5 μM (HSV-1, HSV-2); 15 μM (EBV)), and BrPAA (IC_{50} 5-6 μM (HSV-1, HSV-2); 25 μM (EBV)). The more active compounds had slightly less absolute potency than PAA or PFA (IC_{50} 1-2 μM (HSV-1, HSV-2, EBV)) but showed better selectivity relative to human α-DNA polymerase. None of the compounds tested significantly inhibited human β and γ DNA polymerases. EBV DNA polymerase was somewhat less sensitive to **2b**, **2d** and **2f** than the HSV enzymes, in contrast to the parent compound (PAA) and PFA, for which all three viral polymerases had similar IC_{50} values.

The corresponding dihalo PAA group **2c**, **2e** and **2g** showed substantially less inhibitory activity ($IC_{50} \geq 100$ μM), the most active member of the group being F_2PAA which was 5-10x less effective than FPAA. Among the three mixed dihalo PAA compounds (**2h-2j**), and the three α-halo, α-methyl PAA compounds (**2l-2n**), only FBrPAA had IC_{50} values < 100 μM for the viral enzymes (IC_{50} 37-94 μM). As expected, the 12 triethyl esters of the compounds tested showed no inhibitory activity in the various polymerase screens (data not shown in Table).

Inhibition data for XYPAA samples tested at a single concentration of 100 μM differentiated some of the compounds having an IC_{50} > 100 μM (26). The results indicate that EBV DNA polymerase, but not the HSV-1 and HSV-2 enzymes, had some sensitivity to the α-methyl α-halo PAA derivatives **2l-2m** (40-50% inhibition) and to Cl_2PAA, Br_2PAA and ClBrPAA (30-40% inhibition) (Table II).

None of the corresponding XYMDP salts (**3b-3j**) had IC_{50} < 100 μM with the three viral enzymes tested. The compounds Cl_2MDP, Br_2MDP, and ClMDP were reported to show activity against RNA polymerase from influenza virus A, while PAA was a less effective inhibitor (22,23).

Table I. IC_{50} values (μM) for DNA Polymerase Inhibitors[a]

Compound[b]		Viral HSV-1	Viral HSV-2	Viral EBV	Human α	Human β,γ
PFA,	**1**	1.1	1.1	1.2	15	>300
PAA,	**2a**	1.3	1.2	1.6	21	>300
FPAA,	**2b**	2.5	3.8	14	>100	>300
ClPAA,	**2d**	3.2	5.4	14.5	80	>300
BrPAA,	**2f**	6	5.4	25	>100	>300
F_2PAA,	**2c**	18	30	65	>100	>300
Cl_2PAA,	**2e**	>100	>100	>100	>100	>300
Br_2PAA,	**2g**	>100	>100	>100	>100	>300
FClPAA,	**2h**	>100	>100	>100	>100	>300
FBrPAA,	**2i**	37	70	94	>100	>300
ClBrPAA,	**2j**	>100	>100	>100	>100	>300
CH_3FPAA,	**2l**	>100	>100	>100	>100	>300
CH_3ClPAA,	**2m**	>100	>100	>100	>100	>300
CH_3BrPAA,	**2n**	>100	>100	>100	>100	>300
XYMDP,	**3b-3j**	>100	>100	>100	>100	>300

[a]Reproduced with permission from ref. 26. Copyright 1987 Pergamon.
[b]DCHA salts.

Table II. Percent Inhibition of DNA Polymerases by α-Halo Phosphonoacetates (100 μM)[a]

Compound		Viral HSV-1	Viral HSV-2	Viral EBV	Human α
FPAA,	**2b**	100	100	74	11
ClPAA,	**2d**	96	96	66	45
BrPAA,	**2f**	87	88	85	38
F_2PAA,	**2c**	62	65	77	16
Cl_2PAA,	**2e**	0	0	32	0
Br_2PAA,	**2g**	4.5	4	34	10
FBrPAA,	**2i**	71	70	43	40
ClBrPAA,	**2j**	0	0	33	16
CH_3FPAA,	**2l**	0	0	42	2
CH_3ClPAA,	**2m**	0	0	44	2
CH_3BrPAA,	**2n**	0	0	50	9

[a]Results for compounds tested as both DCHA and Pyr salts are averaged.

Viral Inhibition Results. Table III presents in vitro data (IC_{50} for plaque reduction, and Viral Ratings) for several XYPAA derivatives, PAA and PFA as inhibitors of HSV-1 (F) and two HSV-2 variants (G and Lovelace). Also included are data for an RNA virus (Influenza A/Japan). Data are omitted for the remaining compounds from Table I, all of which had Viral Ratings < 0.4-0.5 for all virus referred to in the Table. The HSV results manifest an approximate correlation with the HSV DNA polymerase activities (Table I). FPAA (data for two different samples) is seen to be midway between PFA and PAA in effectiveness, although another group found it to be somewhat less active than PAA in an HSV-2 plaque reduction assay (12).

Enantiomers of drugs may differ in their metabolism, pharmacokinetics and toxicity. Although PFA and PAA are achiral molecules, α-monosubstituted PAA derivatives, and PAA derivatives α-disubstituted with unlike substituents, possess a chiral center. However, all our current data are for the racemates. In some instances (Tables I, III), racemic FPAA compares favorably in potency with PAA. It would be of interest to determine the relative contributions made by the different FPAA enantiomers to, e.g., HSV vs. human α DNA polymerase inhibition, or to the different parameters (39) which enter into calculation of the Viral Rating.

Table III. Anti-Viral Activities of Some α-Halo Phosphonoacetates in vitro[a]

Compound		VR/IC_{50}[b,c]			
		Influenza A/Japan	HSV-1 (F)	HSV-2 (G)	HSV-2 (Lovelace)
PAA,	**2a**		0.6/190 μM	0.5/190 μM	0.8/190 μM
PFA,	1		0.6/52 μM	0.8/165 μM	0.8/165 μM
ClPAA,	**2d**	0.4/320 μM			
FPAA,	**2b**		0.6/130 μM	0.5/130 μM	0.8/150 μM
Cl_2PAA,	**2e**	0.4/320 μM			
BrPAA,	**2f**		0.4/280 μM		0.6/890 μM
ClBrPAA,	**2j**		0.4/790 μM	0.4/790 μM	

[a]Data obtained at Department of Antimicrobial Research, Syntex Research, Mountain View, CA. [b]VR, the Viral Rating index for the compound, is defined as: < 0.5, insignificant activity; 0.5-0.9, marginal-moderate activity. IC_{50} is here defined as inhibitor concentration giving 50% plaque reduction. For details, and information on assay protocols, see ref. 39. [c]Data for **2b** are averages for separate experiments with two different salts: IC_{50} values are ± 20 μM, VR values are ± 0.1.

HIV-1 Reverse Transcriptase Inhibition Studies.

Effect of Template. The template dependence of HIV-1 reverse transcriptase inhibition (^{3}H-dGTP or ^{3}H-TTP incorporation) by PFA and PAA was determined using poly rA and poly rC. The results are given in Table IV. The IC_{50} values for PFA confirm it to be an active inhibitor, but differed by a factor of five depending on the template used in the assay, poly rA producing the lower of the two values. The same pattern was observed for PAA: with poly rA an IC_{50} of 240 μM corresponding to weak inhibition was found, whereas with poly rC the IC_{50} was at least 60% larger (too large to be measured exactly under the conditions used). Comparison with the data obtained for PFA and PAA inhibition using an activated DNA template (Table V), indicates that the poly rC and 'natural' templates give similar IC_{50} values.

α-Oxo Phosphonates. Inhibition assay results for PFA, PAA, **4**, **5** and pyrophosphate as inhibitors of human α, β and γ DNA polymerases, and of HIV-1 reverse transcriptase are summarized in Table V. Activated DNA template was used in all experiments reported in this Table.

Pyrophosphate (PPi), the nominal reference structure for the analogues, was not inhibitory (IC_{50} < 400 μM). PFA selectively inhibited HIV-1 reverse transcriptase with an IC_{50} value (0.6 μM), almost 50x lower than its IC_{50} for human α DNA polymerase, and at least 10^3 times lower than its IC_{50} values for human β and γ DNA polymerase. The α-oxo phosphonate **4** had an IC_{50} of about 20 μM under the same conditions. When tested under the same standard assay conditions, '**5**' showed an apparent IC_{50} about 6x less than that of **4**. However, **5** was found to react with one or more assay components. These were identified in ^{31}P NMR studies as Tris buffer and dithiothreitol (DTT) (Levy, J.N.; unpublished). Under the same conditions, similar reactions of **4** were not detected.

Table IV. Template Effect on HIV-1 Reverse Transcriptase Inhibition[a]

	IC_{50} (μM)	
Compound	poly rA[b]	poly rC[b]
PFA	0.08	0.4
PAA	240	>400

[a]Assay conditions as described in Biochemical Aspects section on standard Reverse Transcriptase assays, except for template and $[MgCl_2]$ which here = 6 mM. [b]0.5 A_{260} units/mL.

Table V. Inhibition of HIV-1 Reverse Transcriptase by Pyrophosphate Analogues[a]

Compound	Polymerase	IC_{50} (μM)	% Inhibition at 400 μM
PPi	α	> 400	N.D.
	β	> 400	N.D.
	γ	> 400	N.D.
	HIV	> 400	N.D.
PFA, **1**	α	25 ± 3	N.D.
	β	> 400	15 ± 7
	γ	> 400	40 ± 3
	HIV	0.55 ± 0.07	N.D.
PAA, **2a**	α	19.6 ± 6	N.D.
	β	> 400	42 ± 9
	γ	> 400	36 ± 4
	HIV	> 400	43 ± 5
COMDP, **4**	α	> 400	21.5 ± 9.3
	β	> 400	38.4 ± 5.4
	γ	> 400	32.5 ± 3.3
	HIV	20.1 ± 12	N.D.
'COPAA, **5**'	α	210 ± 33	71 ± 2
	β	> 400	25 ± 10
	γ	> 400	5.2 ± 3.7
	HIV	130 ± 50	66 ± 0.7

[a]Standard assay conditions as described in Biochemical Aspects section.

The assay mixture was then modified to replace the Tris with a demonstrably nonreactive buffer (Hepes), with DTT omitted (Biochemical Aspects section). This assay system was used to determine IC_{50} values for 1, 4 and 5 over a pH range of 6.5-8.2 (Table VI). In these experiments poly rA was used as template to maximize sensitivity. Interpolation of the data for 1 in Table VI allows comparison of its IC_{50} value in the modified and standard assay mixtures at pH 8.0 with identical template (poly rA). The difference approaches the limit of error in the two experiments, indicating that the buffer substitution and omission of DTT did not have a drastic effect on the observed inhibition at this pH. COMDP 4 is estimated to be about 20x more active under the modified conditions (Table VI) than when assayed with the standard mixture using an activated DNA template (Table V). Due to its reactivity in the standard assay, the same comparison cannot be made for COPAA 5, but it is seen to be moderately active at pH 8.0, albeit considerably less active than 1 or 4.

The data in Table VI reveal a striking dependence of IC_{50} on pH, with all three inhibitors exhibiting lower values at higher pH. The variation in IC_{50} ranges from 16-17x for the tribasic 1 and 5, to 200x for the tetrabasic 4. An obvious inference from these results is the need for caution in extrapolating reverse transcriptase inhibition assay data (pH 8) to physiological conditions (pH 7) for ionic inhibitors whose charge varies with pH.

A detailed discussion of the ketone/hydrate and anionic acid-base equilibria relevant to 4 and 5 in aqueous solutions will not be attempted here, but qualitatively the parallel activity trends for 1, 4, and 5 appear inconsistent with the hydrate form of 5 (or of 4) being the most important inhibitory species over the pH interval examined.

Table VI. pH Dependence of HIV-1 Reverse Transcriptase Inhibition by Pyrophosphate Analogues[a]

	IC_{50} (μM)		
pH	PFA, 1	COPAA, 5	COMDP, 4
6.50	2.15 ± 0.14	510 ± 91	144 ± 35
6.83	0.616 ± 0.013	154 ± 39	34.0 ± 1.9
7.26	0.303 ± 0.079	74.5 ± 9.9	10.3 ± 0.79
7.74	0.207 ± 0.037	32.4 ± 3.8	1.97 ± 1.2
8.20	0.130 ± 0.072	32.5 ± 4.2	0.719 ± 0.021

[a]Modified assay conditions as described in Biochemical Aspects section.

Nucleotide Halophosphonate Analogues

The mechanism of PFA and PAA resistance by HSV-1 is related to induction of altered HSV-1 DNA polymerases in infected cells (36). It has been suggested that for future treatment of herpes-associated diseases, agents exerting their actions independently should be developed and used in combination to circumvent drug resistance (36). We have recently prepared examples of several nucleotide analogues containing phosphonoacetate moieties (e.g. 13, 14), an anti-herpetic nucleoside (ACV as monophosphate) and an anti-herpetic phosphonoacetate (McKenna, C.E.; Harutunian, V., in press) (Scheme 6). The concept underlying "hybrid" drugs is that a synergistic or other enhanced effect might be obtained by combining a PAA-like inhibitor and a nucleoside drug such as ACV or AZT into a single nucleotide analogue. Such combinations yield a matrix larger than the sum of its components, e.g., 3 PAA analogues and 3 nucleosides provide 9 hybrid pairs. The examples given here were synthesized in adequate yield using a modified dicyclohexylcarbodiimide (DCC) coupling method. While this work was in progress, synthesis and anti-viral data for a series of 2'-deoxyuridine and related pyrimidine nucleoside and acyclonucleoside esters of PFA and PAA were reported (40-41). These derivatives, referred to as 'combined prodrugs', are diphosphate rather than triphosphate nucleotide analogues. Evaluation of their antiviral activity with several herpesviruses produced no evidence for a synergistic action of their combined drug moieties.

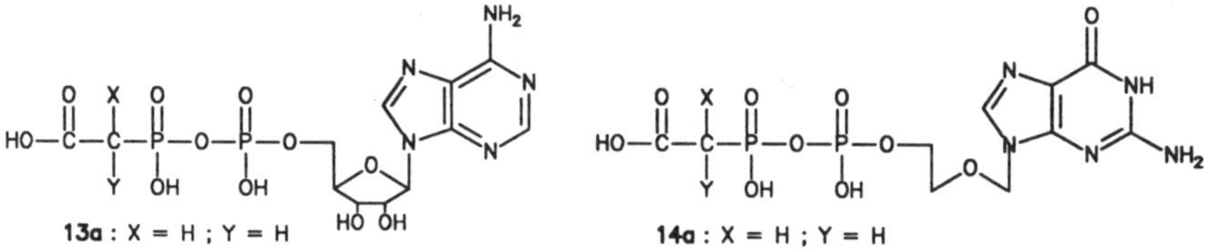

13a : X = H ; Y = H
13b : X = H ; Y = F
13c : X = H ; Y = Cl

14a : X = H ; Y = H
14b : X = H ; Y = Cl

SCHEME 6

Conclusion

α-Halogenation creates a set of PAA congeners of varying polarity with some steric perturbation. These compounds display a significant range of activity as inhibitors of HSV and EBV DNA polymerases. FPAA is the most active inhibitor in this group and also has the highest activity against HSV plaque formation in vitro. Corresponding α-halo MDP derivatives are uniformly inactive as HSV and EBV DNA polymerase inhibitors. Although PFA and PAA have similar IC_{50} values for the herpesvirus DNA polymerases tested, only PFA significantly inhibits HIV-1 reverse transcriptase. The presence of an α-oxo group in PAA or MDP correlates with HIV-1 reverse transcriptase inhibition activity, the MDP derivative 4 having the lower IC_{50}. Hydration of the α-keto groups in 4 and 5 is pH dependent, increasing at lower pH, whereas HIV-1 reverse transcriptase inhibition by 4 and 5 decreases significantly as the assay pH is lowered from 8.2 to 6.5. The pH dependence of HIV-1 reverse transcriptase inhibition by PFA shows a similar trend. This result may be significant in assessing differences between inhibition assays of isolated reverse transcriptase and of viral replication, performed at dissimilar pH. The specificity differences observed for PFA, PAA, MDP and their α-halo and α-oxo derivatives (and for nucleotide inhibitors generally, see elsewhere in this volume and, e.g. (42)) are largely unexplained pending detailed molecular information on substrate and product binding sites of viral polymerases. As progress in this area begins to be made (43,44), a detailed basis for truly rational design of anti-viral agents is likely to emerge.

Acknowledgments

We thank Drs. Anne Bodner and Robert C.Y. Ting (BioTech Research Laboratories) for HIV-1 reverse transcriptase and Dr. Thomas Matthews (Syntex Research) for supplying the in vitro virus inhibition data presented in Table III. Several α-halo methanediphosphonates were prepared at USC by J.-P. Bongartz and P. Pham. This research was supported by NIH grants AI-21871, AI-25697 and CA-44358. J. N. Levy was a 1987-89 University of California University-wide Taskforce on AIDS Postdoctoral Fellow.

Literature Cited

1. Reines, E.D.; Gross, P.A. *Med. Clin. North. Am.* 1988, **72**, 691-715.
2. Cheng, Y.-C.; Ostrander, M.; Derse, D.; Chen, J.-X. *Nucleoside Analogs* 1979, **20**, 319-335.
3. Cheng, Y.-C.; Dutschman G. E.; Bastow, K. F.; Sarngadharan, M. G.; Ting, R.Y.C. *J. Biol. Chem.* 1987, **262**, 2187-2189.
4. Derse D.; Cheng, Y.-C.; Furman, P.A.; St. Clair, M.H.; Elion, G.B. *J. Biol. Chem.* 1981, **256**, 11447-11451.
5. St. Clair, M.H.; Miller, W.H.; Miller, R.L.; Lambe, C.U.; Furman, P.A. *Antimicrob. Agents Chemother.* 1984, **25**, 191-194.
6. Bone, R.; Cheng, Y.-C.; Wolfenden, R. *J. Biol. Chem.* 1986, **261**, 16410-16413.
7. Wahren, B.; Larsson, A.; Ruden, U.; Sundqvist, A.; Solver, E. *Antimicrob. Agents Chemother.* 1987, **31**, 317-320.
8. Christophers, J.; Sutton, R.N. *Antimicrob. Agents Chemother.* 1987, **20**, 389.
9. Lin, J.-C.; DeClercq, E.; Pagano, J.S. *Antimicrob. Agents Chemother.* 1987, **31**, 1431-1433.
10. Robins, R.K. *Pharm. Res.* 1984, 11-18.
11. Prisbe, E.J.; Martin, J.C.; McGee, D.P.F.; Barker, M.F.; Smee, D.F.; Duke, A.E.; Matthews, T.R.; Verheyden, J.P.H. *J. Med. Chem.* 1986, **29**, 671-675.
12. Blackburn, G.M.; Perree, T.D.; Rashid, A.; Bisbal, C.; Lebleu, B. *Chemica Scripta* 1986, **26**, 21-24.
13. Smith, C.C.; Aurelian, L.; Reddy, M.P.; Miller, P.S.; Ts'o, P.O.P. *Proc. Nat. Acad. Sci. USA* 1986, **83**, 2787-2791.
14. Boezi, J.A. *Pharmac. Ther.* 1979, **4**, 231-243.
15. Oberg, B. *Pharmac. Ther.* 1983, **19**, 387-415.
16. Herrin, T.R.; Fairgrieve, J.S.; Bower, R.R.; Shipkowitz, N.L.; Mao, J.C.-H. *J. Med. Chem.* 1977, **20**, 660-663.
17. Noren, J.O.; Helgstrand, E.; Johansson, N.G.; Misiorny, A.; Stening, G. *J. Med. Chem.* 1983, **26**, 264-270.
18. Eriksson, B.; Larsson, A.; Helgstrand, E.; Johansson, N.-G.; Oberg, B. *Biochim. Biophys. Acta* 1980, **607**, 53-64.
19. Eriksson, P.; Oberg, B.; Wahren, B. *Biochim. Biophys. Acta* 1982, **696**, 115-123.
20. Mao, J.C.-H.; Otis, E.; Von Esch, A.M.; Herrin, T.R.; Fairgrieve, J.S.; Shipkowitz, N.L.; Duff, R.G. *Antimicrob. Agents Chemother.* 1985, **27**, 197-202.

21. McKenna, C.E.; Khawli, L.A. *Phosphorus Sulfur* 1984, **18**, 483.
22. Hutchinson, D.W.; Naylor, M. *IRCS Med. Sci.* 1985, **13**, 1023.
23. Hutchinson, D.W.; Naylor, M.; Semple, G. *Chemica Scripta* 1986, **26**, 91-95.
24. Vrang, L.; Oberg, B. *Antimicrob. Agents Chemother.* 1986, **29**, 867-872.
25. McKenna, C.E.; Khawli, L.A. *J. Org. Chem.* 1986, **51**, 5467-5471.
26. McKenna, C.E.; Khawli, L.A.; Bapat, A.; Harutunian, V.; Cheng, Y.-C. *Biochem. Pharm.* 1987, **36**, 3103-3106.
27. McKenna, C.E.; Khawli, L.A.; Bongartz, J.P.; Pham, P.; Ahmad, W.Y.; *Phosphorus Sulfur* 1988, **37**, 1-12.
28. McKenna, C.E.; Levy, J.N. *J.C.S. Chem. Com.* 1989, 246.
29. Kreutzkamp, N.; Mengel, W. *Arch. Pharm.* 1967, **300**, 389-392.
30. Quimby, O.T.; Prentice, J.B.; Nicholson, D.A. *J. Org. Chem.* 1967, **32**, 4111-4114.
31. Dox, A.W. *Org. Synth. Coll.* 1944, 1, 266-269.
32. Martin, M.G.; Ganem, B. *Tetrahedron Lett.* 1984, **25**, 251-254.
33. Regitz, M.; Anschutz, W.; Liedhegener, A. *Chem. Ber.* 1968, **101**, 3734-3743.
34. McKenna, C.E.; Schmidhauser, J. *J. Chem. Soc. Chem. Comm.* 1979, 739.
35. Baril, E.; Mitchner, J.; Lee, L.; Baril, B. *Nucleic Acids Res.* 1977, **4**, 2641-2656.
36. Derse, D.; Barstow, K.F.; Cheng, Y.-C. *J. Biol. Chem.* 1982, **257**, 10251-10260.
37. Tan, R.S.; Datta, A.K.; Cheng, Y.-C. *J. Virol.* 1982, **44**, 893-899.
38. Fisher, P.A.; Wang, T.F.-S.; Korn, D. *J. Biol. Chem.* 1979, **254**, 6128-6133.
39. Sidwell, R.W.; Huffman, J.H. *App. Microbiol.* 1971, **22**, 797-801.
40. Griengl, H.; Hayden, W.; Penn, G.; De Clercq, E.; Rosenwirth, B. *J. Med. Chem.* 1988, **31**, 1831-1839.
41. Lambert, R.W.; Martin, J.A.; Thomas, G.J.; Duncan, I.B.; Hall, M.J.; Heimer, E.P. *J. Med. Chem.* 1989, **32**, 367-374.
42. De Clercq, E. *Biochem. Pharm.* 1988, **37**, 1789-1790.
43. Wang, S.-F.; Wong, S.W.; Korn, D. *FASEB J.* 1989, **3**, 14-21.
44. Gibbs, J.S.; Chiou, H.C.; Bastow J.D.; Cheng, Y.-C.; Corn, D.M. *Proc. Natl. Acad. Sci. USA* 1988, **85**, 6672-6676.

RECEIVED June 15, 1989

Chapter 2

Synthesis of Acyclonucleoside Phosphonates for Evaluation as Antiviral Agents

E. J. Reist[1], P. A. Sturm[1], R. Y. Pong[1], M. J. Tanga[1], and R. W. Sidwell[2]

[1]Life Sciences Division, SRI International, Menlo Park, CA 94025
[2]Utah State University, Logan, UT 84322–0300

Isosteric phosphonate analogs of acyclovir and ganciclovir have been prepared and found to have significant activity against human and murine cytomegalovirus in vitro. The mono ethyl esters also are active, possibly serving as prodrugs for the diphosphonic acids. Higher homologs of the isosteris resulted in significant loss of antiviral activity. The phosphonates appear to have low toxicity and some of them are promising candidates for clinical evaluation.

Chemotherapy of viral infections has lagged significantly behind chemotherapy of bacterial and parasitic infections. One reason for this is that many of the metabolic processes that control the reproduction and growth of bacteria and parasites are unique to the invading organisms and thus offer a means of attack on the invaders by blocking a metabolic transformation that has no close parallel in the metabolism of the host. Viral metabolic processes on the other hand resemble the host processes to a greater extent and a virus in fact will frequently utilize host enzymes to meet its metabolic needs. The net result is that identifying a nontoxic antiviral agent is much more challenging than identifying an effective nontoxic antibacterial or antiparasitic agent.

To this time, there have been only seven compounds that have been licensed by the FDA for treatment of viral diseases (Figure 1). Nucleosides predominate. This is probably a result of fallout from the NCI program of the last 30 years aimed at the synthesis of anticancer agents. There was a heavy nucleoside synthesis component to this program and many of them were available for evaluation in antiviral screens. IUDR (**1**),(1), Trifluorothymidine (**2**),(2) Ara A (**3**),(3) and azidothymidime (AZT)(4) were all originally synthesized on the NCI program. Ribavirin (**5**) is a

0097–6156/89/0401–0017$06.00/0

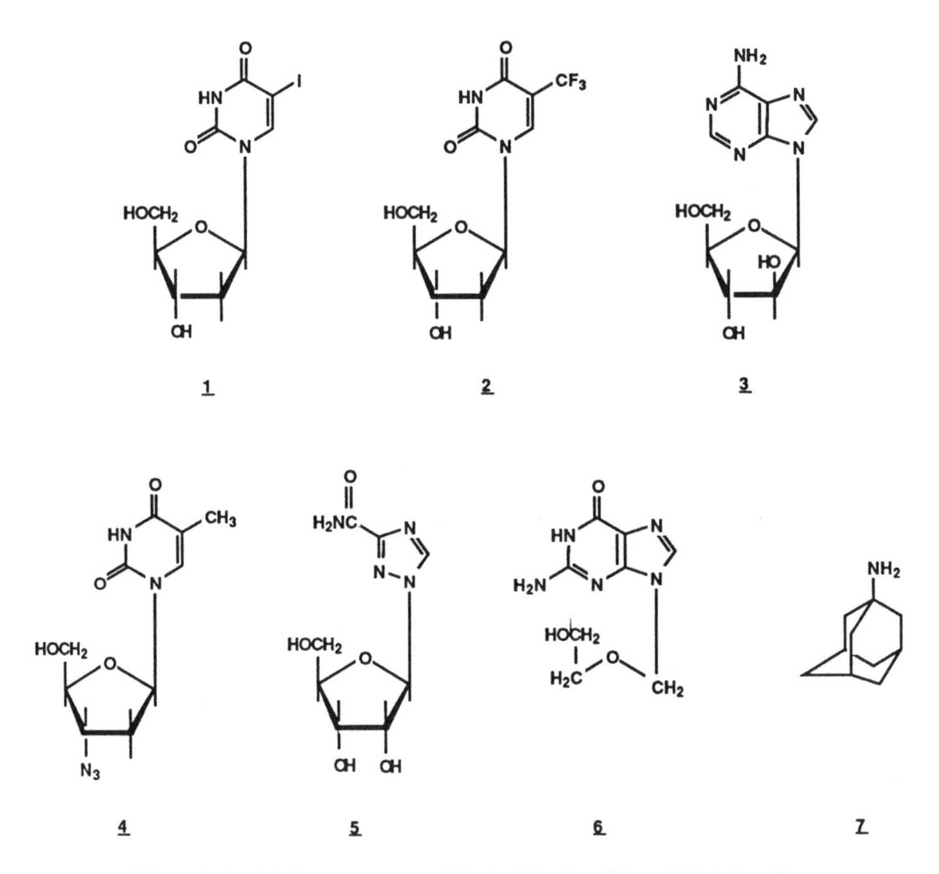

Figure 1. Antiviral agents approved by the Food and Drug Administration.

broad spectrum antiviral that is active against a number of RNA and DNA viruses.(5) Trifluorothymidine,(6) IUDR,(6,7) and Ara A(7) have a much narrower spectrum of activity and their main utility is against the herpes family of viruses, especially herpes simplex virus (HSV). AZT has good activity against human immunodeficiency virus (HIV).(8,9) All five of agents have significant toxicity which limits their overall utility in the clinic.(6,7)

Acyclovir (**6**), the newest member on the list was prepared by Burroughs Wellcome and was found to have outstanding activity against some of the herpes viruses(10)--namely herpes simplex 1 and 2 and varicella zoster. To date it has shown minimal side effects at high concentration in animal studies. The only non-nucleoside on the list--amantadine(**7**)--is one of the oldest antivirals and was prepared by DuPont about 25 years ago and was shown to have activity against influenza A.(11,12) In addition to its antiviral activity, amantadine has shown an effect in Parkinson's disease by affecting dopamine metabolism.(13) This has the unfortunate result that amantadine has side effects due to CNS involvement--insommia, dizziness, mood swings, etc.(14) Thus from this short list of approved compounds, only one--acyclovir--shows good antiviral activity with no significant side effects.

The mechanism by which acyclovir expresses its activity is of interest.(15,16) The herpes virus has a number of unique enzyme systems, among them a viral thymidine kinase (TK). The function of thymidine kinase is to phosphorylate the deoxynucleoside thymidine (**8**) to give thymidine monophosphate (thymidylic acid) (**9**) (Scheme 1). Thymidylic acid is phosphorylated by a thymidylate kinase to give the diphosphate and this is further phosphorylated by another kinase to yield thymidine triphosphate (**10**) which is one of the substrates used by DNA polymerase for the synthesis of DNA. The nucleoside kinases normally have great substrate specificity. Thymidine kinase will only phosphorylate thymidine. Deoxyguanosine kinase will only phosphorylate deoxyguanosine, etc. In the case of the herpes virus thymidine kinase, the substrate specificity is not so great and it can accept acyclovir, a guanine analog, as a substrate to be phosphorylated to the nucleotide (**11**).(7) Subsequently, cellular kinases phosphorylate the acyclovir phosphate (**11**) to the triphosphate (**12**). Acyclovir triphosphate then serves as a substrate for the viral DNA polymerase but is a chain terminator since it does not have the bifunctionality in the side chain that is necessary for DNA chain extension (see **13**). Acyclovir triphosphate is not a substrate for host DNA polymerase, so it has no effect on the host DNA synthesis.

Thus acyclovir manifests its selectivity towards HSV in two ways: (1) It is a substrate for viral thymidine kinase but not for host thymidine kinase. (2) As the triphosphate, it is a substrate for viral DNA polymerase but not for host DNA polymerase. A change in the viral TK can result in resistance of

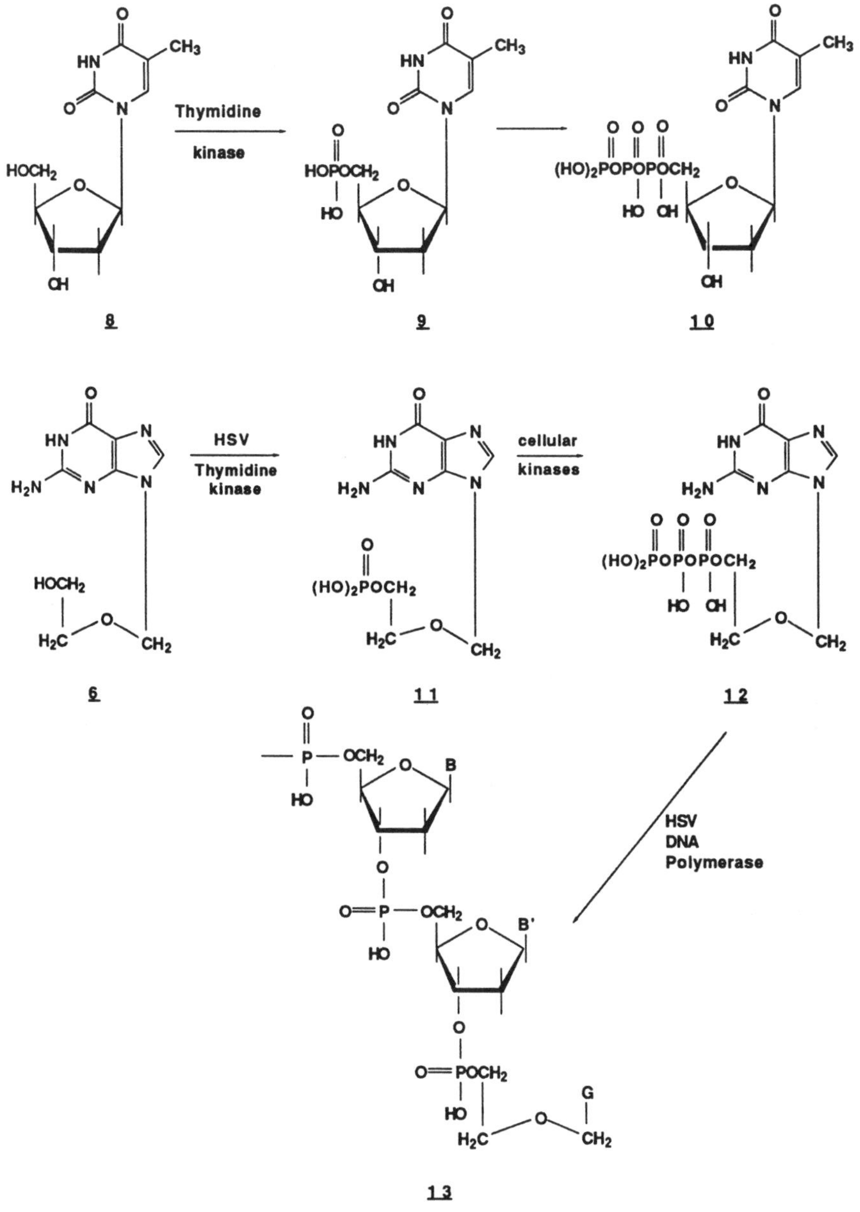

8 → Thymidine kinase → 9 → 10

6 → HSV Thymidine kinase → 11 → cellular kinases → 12

12 → HSV DNA Polymerase → 13

Scheme 1

the herpes virus to acyclovir. This has been demonstrated in the laboratory and herpes virus that does not have its own thymidine kinase is indeed resistant to acyclovir. Fortunately thymidine kinase negative herpes has not developed into a clinical problem at this time, although it is always a threat.

An analog of acyclovir, ganciclovir (**14**) (Figure 2) also has outstanding activity against HSV-1, HSV-2 and varicella zoster.(18) In addition to these viruses, ganciclovir has in vivo activity against cytomegalovirus (CMV), a member of the herpes family that does not have a specific viral thymidine kinase.(19) It is probable that ganciclovir is phosphorylated by the viral thymidine kinase of herpes simplex to the triphosphate. However, due to the extra functionality, it is not necessarily a chain terminator, so its antiviral effect must be somewhat different from that of acylovir. The activity of ganciclovir against CMV is also not fully understood. Since CMV has no viral thymidine kinase the mechanism by which it inhibits the virus is not obvious. It has been speculated that the CMV is able to stimulate the host thymidine kinase to excessive phosphorylation.(20) Possibly in the process, the selectivity of the cellular thymidine kinase is somewhat compromised. It is interesting to note that acyclovir with its monofunctional character must be a DNA chain terminator, has minimal side effects and is essentially nontoxic. Ganciclovir, a close structural relative with its bifunctional character can get incorporated into DNA and has many toxic side effects--neutropenia, leukopenia, testicular atrophy, azospermia, atrophy of the G.I. mucosa, etc.(21) Although ganciclovir is the only drug to show adequate in vivo activity against CMV to date, it has not yet been cleared by the FDA for this purpose because of concern over these side effects.

The acyclovir/ganciclovir story offered an unusual opportunity for analog development to prepare new compounds that maintain the desired antiviral activity while at the same time, the side effects observed for ganciclovir could possibly be decreased or eliminated. The preparation of isosteric phosphonates of ACV and ganciclovir was an especially attractive target for analog development.

As can be seen, acyclovir phosphate (**11**) and the isosteric phosphonate (**15**) are quite similar. Space filling models indicate that both are of very similar bulk. Since acyclovir phosphate can be phosphorylated to the triphosphate (**12**) by cellular kinases, it could be hoped that the isosteric phosphonate (**15**) could also be a substrate for these same kinases to give an analogous triphosphate (**16**) (Scheme 2). Generally speaking, nucleoside phosphates have difficulty crossing the cell membrane to enter a cell because the highly polar nature of the phosphate moiety is not compatible with the lipophilic character of the cell membrane. However, it has been demonstrated(22) that cells that have been infected by virus have a modified cell membrane that is more permeable to a polar molecule, such as a phosphate. In addition, a phosphonate is somewhat less polar than phosphate, so it is possible that a

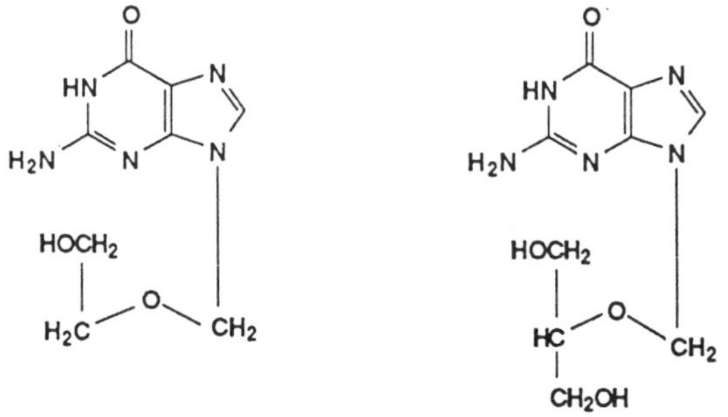

Acyclovir

DHPG, 2'-NDG
BW759U, Biolf 62
Ganciclovir

Figure 2. Acyclic Nucleosides Active Against the Herpes Family of Viruses

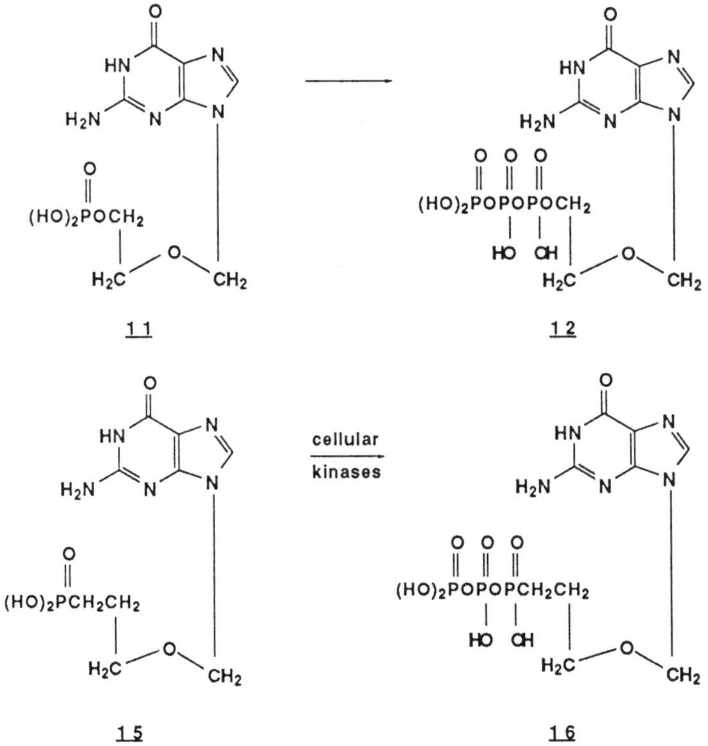

11 → 12

15 —cellular kinases→ 16

Scheme 2

nucleoside phosphonate such as shown here could have some degree of selective absorption into an infected cell and thus be concentrated in the cells that need it.

With this as the rationale, a number of acyclonucleoside phosphonates have been prepared for evaluation as antivirals.

The syntheses started as outlined in Scheme 3 with the reaction of triethylphosphite (**17**) with 1,3-dibromopropane (**18**) to give diethyl 3-bromopropyl phosphonate (**19**) in 71% yield. Replacement of the bromide by acetate was accomplished using sodium acetate in DMF. Acid catalyzed hydrolysis of the O-acetate (**20**) using Dowex 50 (H^+) gave the 3-hydroxypropyl phosphonate (**21**) in 50% overall yield from the bromide. Chloromethylation was accomplished using paraformaldehyde and hydrogen chloride to give the 3-chloromethoxy propyl phosphonate (**22**), suitable for coupling with an appropriate purine. A number of approaches have been used (Scheme 4) to couple the chloromethyl sugar to a guanine analog. The most convenient from the standpoint of ease of handling, solubility characteristics, etc. utilized 2-amino-6-chloropurine (**23**) as starting material. Treatment of this purine with hexamethyl-disilazane gave a di-trimethylsilyl derivative (**24**) which was condensed with the chloromethyl ether (**22**) to give a 47% yield of crystalline 2-amino-6-chloropurine nucleoside phosphonate as the diethyl ester. Treatment of the blocked acyclonucleoside with tetraethylammonium hydroxide and trimethylamine hydrolized the 6-chloro group and gave a 72% yield of crystalline guanine nucleoside phosphonate (**28**) as the diethyl ester. Heating the diethyl ester (**28**) with conc. ammonium hydroxide gave a 65% yield of the monoethyl ester (**30**) that was homogeous on TLC and HPLC with satisfactory UV spectrum for a 9-substituted guanine. An alternative synthesis was developed, starting from guanine (**26**). Silation of (**26**) was accomplished by standard procedures to yield a very labile disilyl guanine derivative (**27**) that was alkylated directly with the chloromethyl ether (**22**). When 1 mole of mercuric cyanide was present,a yield of 35% of 9-substituted guanine (**28**) was obtained with small amounts (≈2%) of 7-substituted isomer (**29**). If mercuric cyanide was omitted during the condensation, yields up to 55% were obtained, however the product obtained contained 10% of (**29**) which could be separated by reverse phase chromatography of the monoesters (**30** and **32**).

Complete deesterification of the phosphonate (**30**) was readily accomplished by treatment of the mono ethyl ester with bromo-trimethylsilane to yield the diacid (**31**) in good yield after purification.

By a similar sequence, starting from 1,7-dibromo heptane (**33**), a heptyl analog (**35**) was prepared (Scheme 5). The rationale to prepare such a compound was based on the idea that such a molecule is very similar in chain length to the triphosphorylated side chain of acyclovir that is believed to be the ultimate active antiviral, although certainly the heptyl phosphonate is significantly different in polar character.

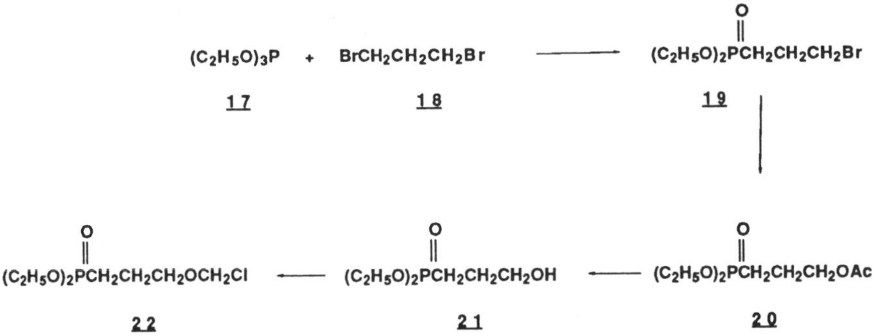

$(C_2H_5O)_3P$ (**17**) + $BrCH_2CH_2CH_2Br$ (**18**) → $(C_2H_5O)_2P(O)CH_2CH_2CH_2Br$ (**19**)

→ $(C_2H_5O)_2P(O)CH_2CH_2CH_2OAc$ (**20**) → $(C_2H_5O)_2P(O)CH_2CH_2CH_2OH$ (**21**) → $(C_2H_5O)_2P(O)CH_2CH_2CH_2OCH_2Cl$ (**22**)

Scheme 3

23 → **24** (TMSNH, TMS) → **25** ($(EtO)_2P(O)(CH_2)_3OCH_2$)

26 → **27** (OTMS, TMSNH, TMS) → **28** ($(EtO)_2P(O)(CH_2)_3OCH_2$) + **29** ($(EtO)_2P(O)(CH_2)_3OCH_2$)

28 → **30** ($EtOP(O)(OH)(CH_2)_3OCH_2$) → **31** ($(HO)_2P(O)(CH_2)_3OCH_2$)

29 → **32** ($EtOP(O)(OH)(CH_2)_3OCH_2$)

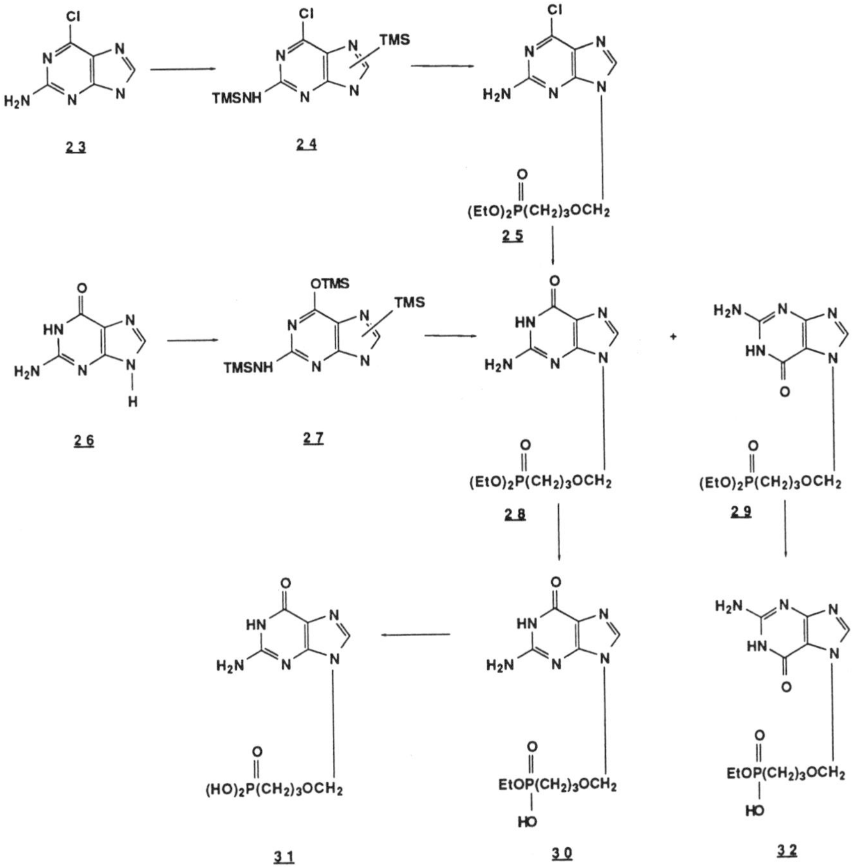

Scheme 4

The preparation of the next higher homolog (**39**) of acyclovir phosphonate was accomplished by the sequence of reactions in Scheme 6. Hydroboration of diethyl 3-butene phosphonate (**36**) gave the 4-hydroxybutyl phosphonate (**37**). Chloromethylation followed by reaction with silated 2-amino-6-chloropurine gave the expected 2-amino-6-chloropurine butyloxymethyl phosphonate (**38**) as the diethyl ester. Selective hydrolysis gave the monoester (**39**).

Bromination of the monoester (**30**) gave the 8-bromo derivative (**40**) in good yield (Scheme 7).

Hydrogenolysis of the 2-amino-6-chloropurine diethyl ester (**25**) gave the 2-aminopurine diethyl ester which was then selectively hydrolized to give the mono ester phosphonate of 2-aminopurine (**42**) (Scheme 7). The rationale for the preparation of this compound is based on the observation that the 2-aminopurine analog of acyclovir is an excellent prodrug for acyclovir and has significantly higher oral bioavailability than acyclovir.(23) It is readily functionalized in vivo by xanthine oxidase to produce acyclovir in situ. However, the 2-aminopurine analog of acyclovir is significantly more toxic than acyclovir, possibly due to cleavage in vivo to 2-aminopurine which is itself quite toxic.

In the above syntheses, the end products are generally the monoethyl esters rather than the diacids. We have observed that the monoethyl ester appeared to have comparable or even better activity than the diacid and seemed to have superior solubility characteristics. A monoester would also be less polar than the diacid hence could possibly penetrate the cell membrane more easily and could conceivably be hydrolized to the diacid by esterases within the cell. Thus it seemed reasonable to evaluate the effect of higher esters and the mono butyl (**50**) ester was prepared by an identical reaction sequence as described for the preparation of the ethyl ester (Scheme 8).

Phosphonate derivatives of ganciclovir were prepared by a sequence of reactions starting from diethyl methylphosphonate (**51**) (Scheme 9). Preparation of the lithium salt, using butyl lithium/cuprous iodide, followed by reaction with allyl bromide gave an 80% yield of diethyl 3-butenyl phosphonate (**36**). Epoxidation of the olefin with meta chloroperbenzoic acid gave a 75% yield of the epoxide (**52**). Acid catalyzed cleavage of the epoxide with glacial acetic acid gave a 50% yield of diethyl 4-acetoxy-3-hydroxy butylphosphonate (**53**). The M/S cracking pattern showed the presence of CH_2OAc, indicating that the desired isomer was formed.

Chloromethylation gave the desired chloromethyl ether (**54**) which was condensed with the di(TMS) derivative of 2-amino-6-chloropurine to give the blocked nucleoside. Treatment with aqueous 1N NaOH gave the monoethyl phosphonate (**55**) in 21% overall yield. The mono ester was also completely deblocked by treatment with bromotrimethyl silane followed by water to give the phosphonic diacid (**56**). Cyclization of the diacid using DCC in

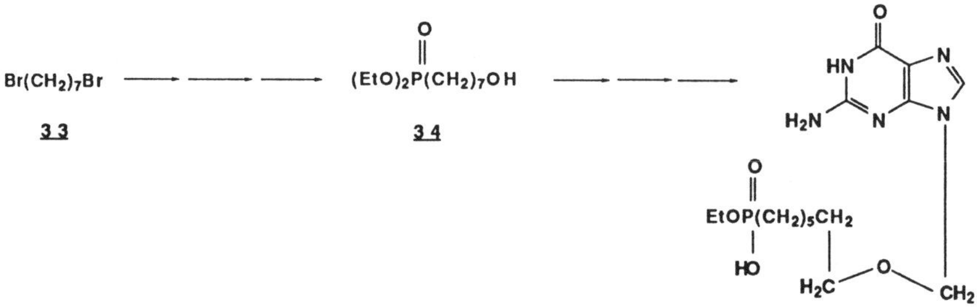

$Br(CH_2)_7Br$ → → → $(EtO)_2P(O)(CH_2)_7OH$ → → → 35

33 34

35: $EtOP(O)(OH)(CH_2)_5CH_2$–H_2C–O–CH_2–guanine (N9)

Scheme 5

$(EtO)_2P(O)CH_2CH_2CH{=}CH_2$ (36) → $(EtO)_2P(O)CH_2CH_2CH_2CH_2OH$ (37)

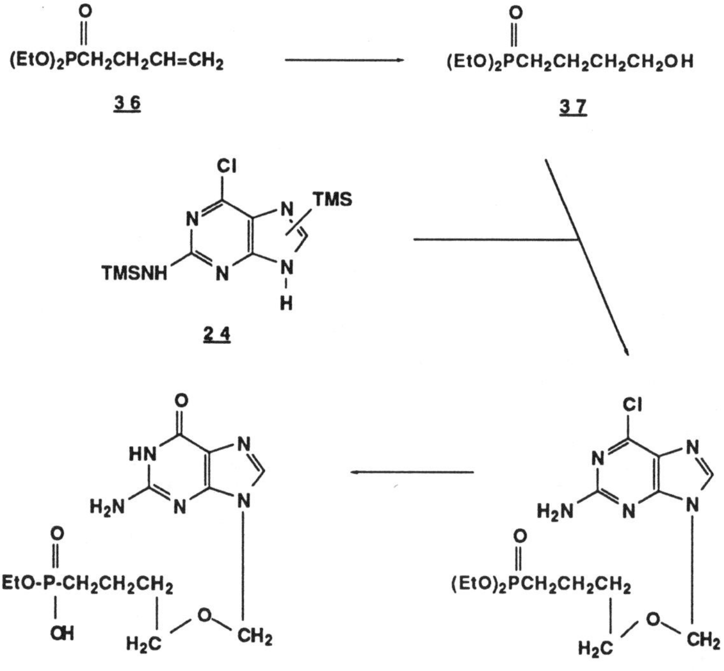

24 (2-TMSNH-6-Cl-purine, N-TMS) + 37 → 38

38: $(EtO)_2P(O)CH_2CH_2CH_2$–H_2C–O–CH_2–(2-amino-6-chloropurine, N9)

38 → 39

39: EtO-P(O)(OH)-$CH_2CH_2CH_2$–H_2C–O–CH_2–guanine (N9)

Scheme 6

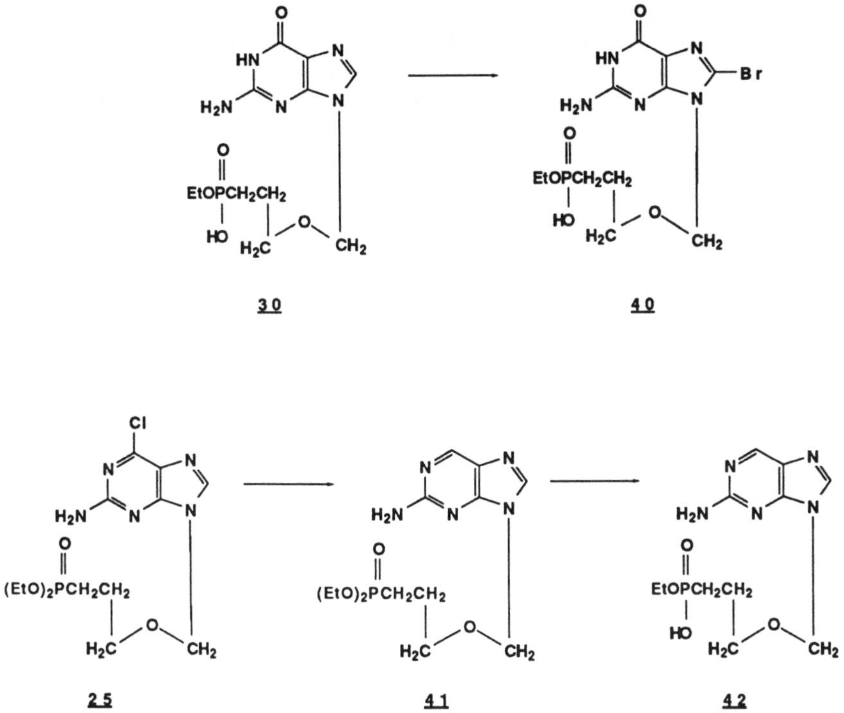

30 → 40

25 → 41 → 42

Scheme 7

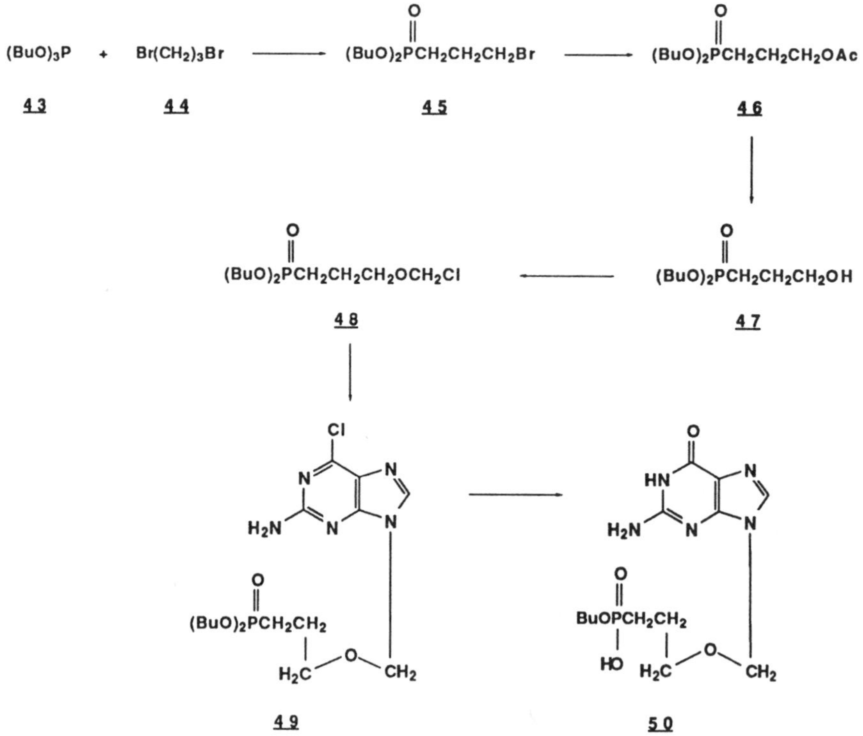

$(BuO)_3P$ + $Br(CH_2)_3Br$ → $(BuO)_2P(O)CH_2CH_2CH_2Br$ → $(BuO)_2P(O)CH_2CH_2CH_2OAc$

43 44 45 46

↓

$(BuO)_2P(O)CH_2CH_2CH_2OCH_2Cl$ ← $(BuO)_2P(O)CH_2CH_2CH_2OH$

48 47

↓

49 → 50

Scheme 8

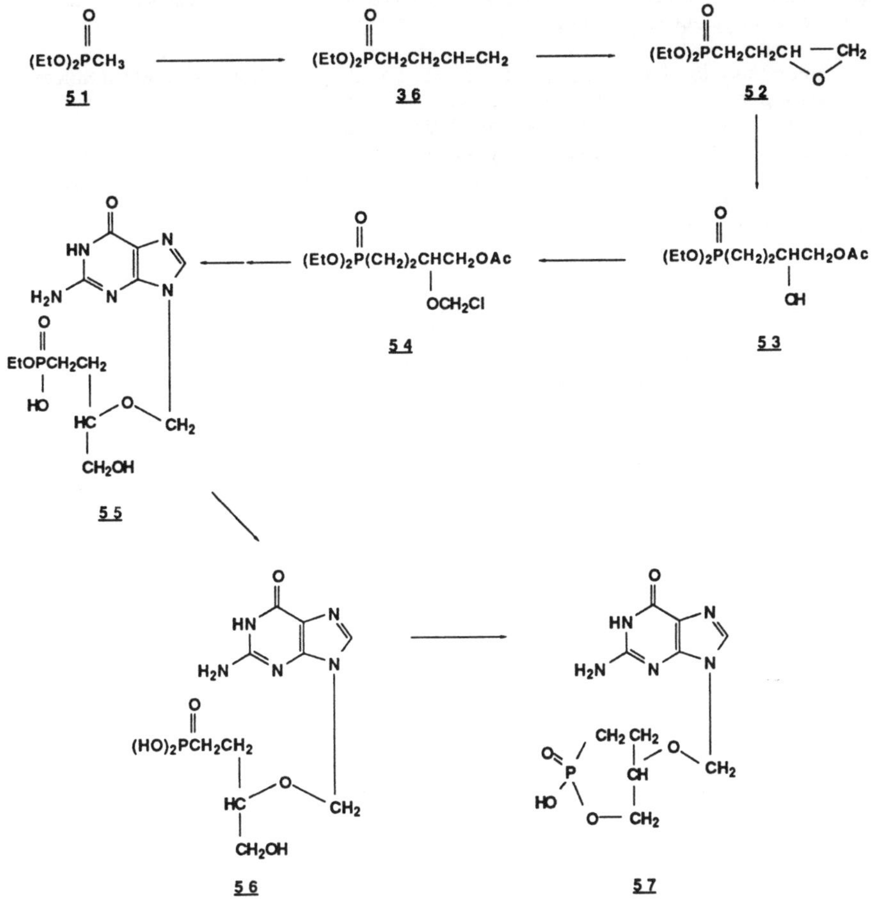
(EtO)2PCH3
51
(EtO)2PCH2CH2CH=CH2
36
(EtO)2PCH2CH2CH —CH2
52
(EtO)2P(CH2)2CHCH2OAc
OH
53
(EtO)2P(CH2)2CHCH2OAc
OCH2Cl
54
HN
H2N
EtOPCH2CH2
HO
HC
CH2
CH2OH
55
(HO)2PCH2CH2
CH2OH
56
CH2 CH2
CH
O— CH2
57

Scheme 9

pyridine gave the cyclic phostonate (**57**) in reasonable yield. The HPLC behavior of the monoester was interesting. It consistantly showed a double peak that was not adequately resolved for separation. The ultraviolet spectrum and analysis were as expected for the assigned product. On complete deesterification, this behavior disappeared and the material was homogeneous in HPLC, UV and analysis were as expected. On preparation of the phostonate, two peaks again appeared, again with satisfactory UV and elemental analysis. We assume this is due to 2 isomeric pairs due to 2 asymetric centers.

The branched methyl analog of ganciclovir phosphonate (**60**) (Scheme 10) was prepared for two reasons: (1) The analogous derivative of ganciclovir showed antiviral activity comparable to ganciclovir for rate of phosphorylation by HSV-1 thymidine kinase(24) thus showing that the methyl group was compatible with the HSV enzyme system. (2) If it showed adequate activity, it would be an interesting monofunctional analog of ganciclovir that cannot propagate the DNA chain.

Opening the epoxide (**52**) using borohydride gave the expected secondary alcohol (**58**). Chloromethylation followed by coupling with 2-amino-6-chloropurine gave after deblocking the deoxyganciclovir phosphonic acid, mono ethyl ester (**60**).

Some of the biological activity that we have obtained is shown in **Table 1**. Acyclovir phosphonate (**31**) and its monoethyl ester (**32**) both show moderate activity against HSV-1 although neither as active as acyclovir or ganciclovir. Surprisingly, (**31**) was inactive against murine cytomegalovirus at doses up to 1000 µg/m although the mono ester showed good activity and a very good therapeutic index.

The heptyl analog (**35**) showed low activity against both HSV-1 a human cytomegalovirus. Likewise the diethyl ester (**28**) and 8-bromo analog (**40**) were essentially inactive. The results obtained for the monobutyl ester (**50**) were somewhat surprising. There was no activity against HSV-1. Although there was activity against HCMV, it was significantly lower than that observed for the mono ethyl ester (**32**). The activity of the ganciclovir analogs prepared so far is consistently low against HSV-1. However some of them show good activity against human cytomegalovirus. The diacid (**56**) shows a therapeutic index of 500 with an ED_{50} of 2 µg/mL, comparable to ganciclovir in activity. The monoethyl ester (**55**) is nearly as effective with a therapeutic index of 258 and an ED_{50} of 5.8 µg/mL. The cyclic phostonate (**57**) is less active with a therapeutic index of 64 and an ED_{50} of 75 µg/mL, while the deoxy analog (**60**) has only low activity with a therapeutic index of 5 and an ED_{50} of 400 µg/mL.

We have done some studies on the toxicity of the monoethyl ester (**30**) of ACV phosphonate in the mouse. An acute dose of 2000 mg/kg in the tail vein of the mouse gave no significant adverse reaction. On chronic dosage once daily over 5 days at 2000 mg/kg

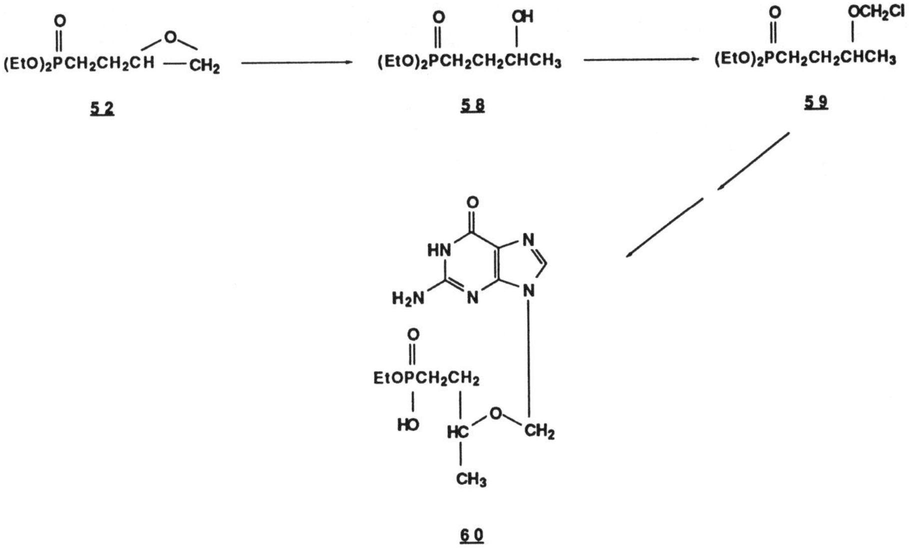

$(EtO)_2P(O)CH_2CH_2CH(-O-)CH_2$ → $(EtO)_2P(O)CH_2CH_2CH(OH)CH_3$ → $(EtO)_2P(O)CH_2CH_2CH(OCH_2Cl)CH_3$

52 58 59

O, HN, N, N, H_2N, N, N

$EtOP(O)(HO)CH_2CH_2$

$HC-O-CH_2$

CH_3

60

Scheme 10

Table 1. Antiviral Activity of Acyclic Phosphonates Against Herpes Viruses

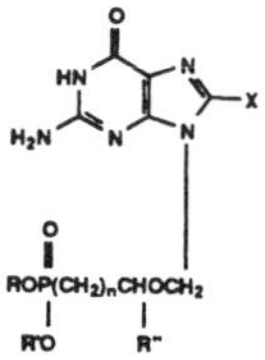

Cpd	X	R	R'	R"	n	HSV-1 V.R.[1]	HSV-1 ED_{50} µg/ml	HSV-1 T.I.[2]	HCMV ED_{50} µg/ml	HCMV T.I.[2]	MCMV V.R.[1]	MCMV ED_{50} µg/ml	MCMV T.I.[2]
31	H	H	H	H	2	0.6	100	>10			0		
30	H	C_2H_5	H	H	2	0.9	110	10	16	94	1.6	10	300
39	H	C_2H_5	H	H	3				>320	N.A.[3]			
35	H	C_2H_5	H	H	6	0.1	>1000[4]		1000[4]	1			
28	H	C_2H_5	C_2H_5	H	2	0.1	1000[4]		1000[4]	N.A.[3]			
40	Br	C_2H_5	H	H	2	0			<500	>2			
50	H	C_4H_9	H	H	2	0			100	15			
55	H	C_2H_5	H	CH_2OH	2	0.3	1000[4]	1	5.8	258			
60	H	C_2H_5	H	CH_3	2	0.1	320	0.3	400	5			
56	H	H	H	CH_2OH	2	0.3	320	3	2	500		1.3	700[5]
57	H	H	CH_2		2	0.1	320	0.1	5	64			
		Acyclovir				1.8	4	375	25	60			
		Ganciclovir				1.8	2	500	2	500			
		DHPG Cyclic Phosphate					10-20		6	20[5]		1.6	75[5]

[1]V.R. is virus rating and represents a measure of antiviral activity.(25) A rating between 0 and 0.5 is low activity. A rating between 0.5 and 1.0 is moderate activity. A rating greater than 1.0 is strong activity.
[2]T.I. is therapeutic index and is the ratio of cytotoxic dose (CD_{50}) to minimum effective dose (ED_{50}).
[3]N.A. is not active.
[4]Top dose tested is 1000 µg/mL.
[5]Data from ref. 26.

and also as 1000 mg/kg, there was 1 death each in 5 mice. At 500 mg/kg there were no toxic effects.

The diacid of ganciclovir phosphonate (**56**) has also been prepared by Prisbe, et al.(26) They also observed that the herpes activity was minimal but that there was good activity against HCMV *in vitro*. **56** also showed excellent activity when applied subcutaneously against MCMV. They reported low toxicity for the compound and believe it is a good candidate for clinical evaluation against HCMV.

Acknowledgment

This work was supported in part by contract NO1-AI-72643 from the National Institute of Allergy and Infectious Diseases.

References

1. Prusoff, W. H. *Biochim. Biphys. Acta*. 1959, *32*, 295.
2. a) Heidelberger, C.; Parsons D.; Remy, D. C. *J. Am. Chem. Soc*. 1962, *84*, 3597.
   b) Ryan, K. J.; Acton, E. M.; Goodman, L. *J. Org. Chem*. 1966, *31*, 1181.
3. a) Lee, W. W.; Benitez, A.; Goodman, L.; Baker, B. R. *J. Am. Chem. Soc*. 1960, *82*, 2648.
   b) Reist, E. J.; Benitez, A.; Goodman, L.; Baker, B. R.; Lee, W. W. *J. Org. Chem*. 1962, *27*, 3274.
4. Horowitz, J. P.; Chua, J.; Noel, M. *J. Org. Chem*. 1964, *29*, 2076.
5. Sidwell, R. W.; Huffman, J. H.; Khare, G. P.; Allen, L. B.; Witkowski, J. T.; Robins, R. K. *Science* 1972, *177*, 705.
6. Nicholson, K. G. *Lancet II* 1984, 617.
7. Nicholson, K. G. *Lancet II* 1984, 503.
8. Mitsuya, H., et al. *Proc. Nat. Acad. Sci. USA* 1985, *82*, 7096.
9. Fischl, M. A., et al. *New Eng. J. Med*. 1987, *317*(4), 185.
10. Collins, P.; Bauer, D. J. *J. Antimicrob. Chemoth*. 1979, *5*, 431.
11. a) Galbraith, A. W.; Oxford, J. S.; Schild, G. C.; Watson, G. I. *Lancet* 1969, 1026.
    b) *Bull. WHO* 1969, *41*, 677.
12. Nicholson, K. G. *Lancet II* 1984, 562.
13. Allen, R. M., et al. *Clin. Neuropharmacol*. 1983, *6*, (Suppl. 1), S64.
14. a) Bryson, Y. J.; Monahan, C.; Pollack, M.; Shields, W. D. *J. Infec. Dis*. 1980, *141*, 543.
    b) Flaherty, J. A.; Bellur, S. N. *J. Clin. Psychiat*. 1981, *42*, 344.
15. Elion, G. B.; Furman, P. A.; Fyfe, J. A.; de Miranda, P.; Beauchamp, L.; Schaeffer, H. J. *Proc. Nat. Acad. Sci USA* 1977, *74*, 5716.
16. Furman, P. A.; de Miranda, P.; St. Clair, M. H.; Elion, G. B. *Antimicrob. Agts. Chemother*. 1981, *20*, 518.
17. Elion, G. B. *Amer. J. Med*. 1982, *73*, 7.

18. a) Smith, K. O.; Galloway, K. S.; Kendall, W. L.; Ogilvie, K. K.; Radatus, B. K. Antimicrob. Agts. Chemother. 1982, 22, 55.
b) Martin, J. C.; Dvorak, C. A.; Smee, D. F.; Matthews, T. R.; Verheyden, J. P. H. J. Med. Chem. 1983, 26, 759.
19. Mar, E.-C.; Cheng, Y.-C.; Huang, E.-S. Antimicrob. Agts. Chemother. 1983, 24, 518.
20. Estes, J. E.; and Herang, E.-S. J. Virol. 1977, 24, 3.
21. Koretz, S. H., et al. New Eng. J. Med. 1986, 314, 801.
22. Carrasco, L. Nature 1978, 272, 694.
23. Krenitsky, T. A.; Hall, W. W.; deMiranda, P.; Beauchamp, L. M.; Schaeffer, H. J.; and Whiteman, P. D. Proc. Nat. Acad. Sci, USA 1984, 81, 3209.
24. Fyfe, J. A.; McKee, S. A.; and Keller, P. M. Mol. Pharmacol. 1983, 24, 316.
25. Sidwell, R. W. Viral Diseases: A Review of Chemotherapy Systems; In "Chemotherapy of Infectious Diseases" (H. Gadebusch, ed.) pp. 31-54, CRC Press, Cleveland.
26. Prisbe, E. J.; Martin, J. C.; McGee, D. P. C.; Barker, M. F.; Smee, D. F.; Duke, A. E.; Matthews, T. R.; Verheyden, J. P. H. J. Med. Chem. 1986, 29, 671.

RECEIVED February 22, 1989

Chapter 3

Structural Requirements for Enzymatic Activation of Acyclonucleotide Analogues

Relationship to Their Mode of Antiherpetic Action

Richard L. Tolman

Merck Sharp & Dohme Research Laboratories, Box 2000, Rahway, NJ 07065

The structure-substrate relationships of the viral specified thymidine kinase (TK) as well as host phosphorylating enzymes (GMP kinase) have been studied and relevant acyclonucleoside conformations amenable to phosphorylation have been proposed based on molecular modelling. The modes of action of other acyclonucleoside antiviral agents, which are not viral DNA polymerase inhibitors, and therefore TK-independent, *e.g.* 2'nor-cGMP, have also been discussed.

The most safe and effective agents of the present generation of antivirals for members of the group of herpes viruses are guanine acyclonucleosides. These agents have been shown(1,2) to ultimately inhibit the viral replication process as their triphosphate derivatives.

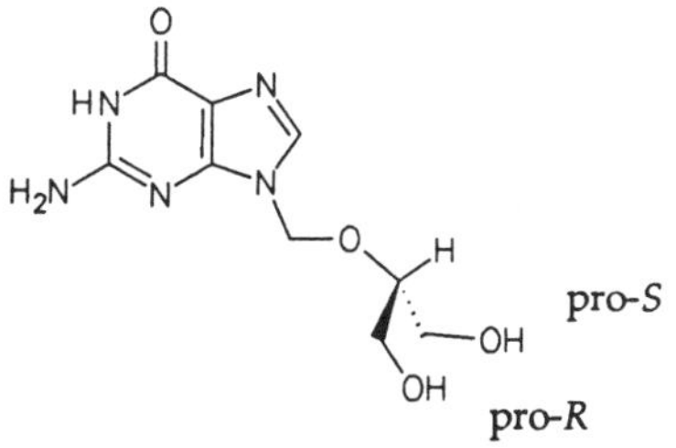

O HN N H_2N N N O H pro-*S* OH OH pro-*R*

Ganciclovir

O HN CH_3 O N O OH OH

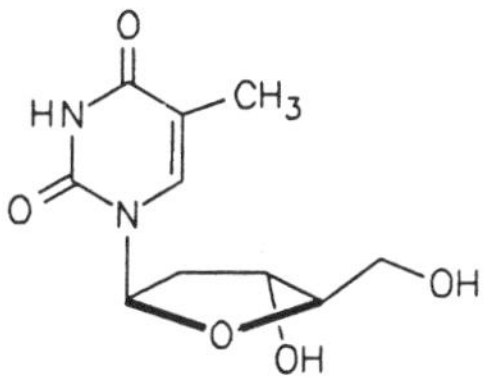

Thymidine

It is a viral-specified enzyme, thymidine kinase (TK), which accomplishes the first phosphorylation(3) of guanine acyclonucleosides, such as acyclovir (ACV), to monophosphate thereby accounting for the great selectivity of their antiviral action. Our initial interest in this area was in understanding the manner of this remarkable enzymatic event in accepting a guanine derivative as a surrogate for the natural substrate thymidine. The structure-substrate relationships for acyclonucleoside derivatives were examined for herpes (HSV-1) thymidine kinase as well as for the host enzyme GMP kinase, which converts monophosphate to diphosphate(4). There are many

0097–6156/89/0401–0035$06.00/0

enzymes(5) which convert acyclonucleoside diphosphate to triphosphate, the final agent which inhibits viral DNA polymerase and thereby viral replication (Fig. 1). Molecular modelling was found to be a source of stimulation toward the synthesis of new chemical entities and to provide plausible scenarios for enzymatic phosphorylation.

Molecular modelling studies are arguably less relevant in their application to substrate studies as there are two components in substrate activity, binding and efficacy. It is the binding component which is more easily dealt with in the absence of an X-ray structure of the enzyme.

HSV Thymidine Kinase

Keller *et al.*(6) have examined in a cursory fashion the substrate requirements of herpes thymidine kinase (HSV-TK) particularly with regard to guanine derivatives. The structural requirements of pyrimidine nucleoside analogs have been examined(7-10) for substrate activity. It has been demonstrated that 5-substitution of the pyrimidine ring is important for substrate activity. Thymidine is a much better substrate than uridine which has hydrogen as a 5-substituent. The ability of other bulkier and more hydrophobic 5-substituents to function even more effectively than thymidine as substrates has led to the thesis that the binding site of TK which accommodates the 5-methyl of thymidine is in fact a long cylindrical hydrophobic pocket(7). 2′-Deoxy-cytidine is also a quite a good substrate(one-twentieth as effectively phosphorylated as thymidine, 11) despite the fact that it has hydrogen at C5.

The phosphorylation of ganciclovir (GCV) by HSV-TK has been shown to be stereospecific for the pro-*S* hydroxyl(12) as evidenced by the fact that enzymatically-prepared GCV-MP is rapidly converted by GMP kinase to GCV-DP, whereas the chemically-prepared MP (racemic) is only 50% converted to DP under the same conditions.

Molecular Models. A statement of Garland Marshall's(13) describes the promise of molecular modelling for studies of the type of the thymidine kinase structure-substrate study: "Without detailed information about the three-dimensional nature of the receptor, conventional physicochemically based approaches are not possible. One can only attempt to deduce an operational model of the receptor that gives a consistent explanation of the known data and, ideally, provides predictive value when considering new compounds for synthesis and biological testing."

Using a modified MM2 force field(14), three dimensional energy maps were generated for thymidine and ganciclovir (Figs 2 & 3) using two side-chain dihedral angles(15). The conformationally-constrained thymidine was seen to have a number of low energy wells corresponding to anti-conformers with gauche- or gauche+ 5′-hydroxyls, whereas permutations of GCV's side chain position produced little variation in energy (total energy range = 10 kcal). The C8-N9-C1′-O torsion angles from the X-ray structures of ACV(16-18) and GCV(19,15) (-91° and -110° respectively) fell in a low energy portion of the contour plot, but were not the low*est* energy structures in that energy well.

The key to understanding how HSV-TK could accept a guanine acyclonucleoside as a substrate seemed to be in the positioning of the purine on the active site of the enzyme. The effect of some purine substituent changes upon TK substrate ability have been examined by others(20,6). Table I shows the relative substrate ability of various substituted guanine derivatives within any of three side-chain series. Any change of the 6-oxo function abolished substrate activity [the substrate activity of 2′-

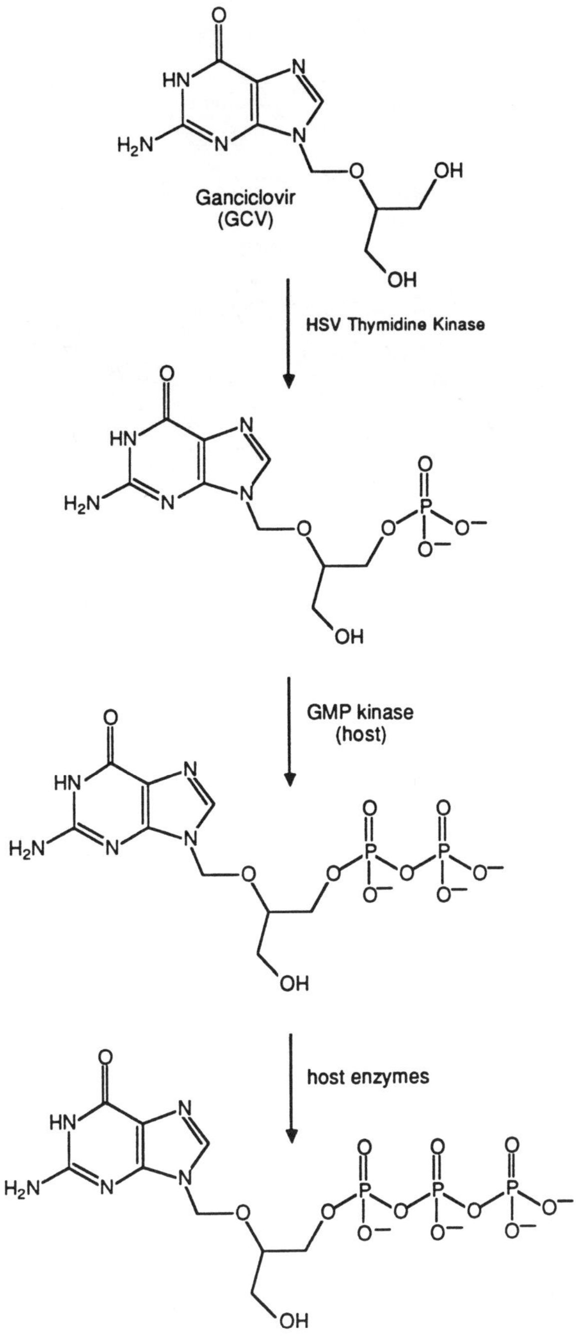

Ganciclovir
(GCV)

HSV Thymidine Kinase

GMP kinase
(host)

host enzymes

Figure 1. Viral Activation of Acyclonucleosides

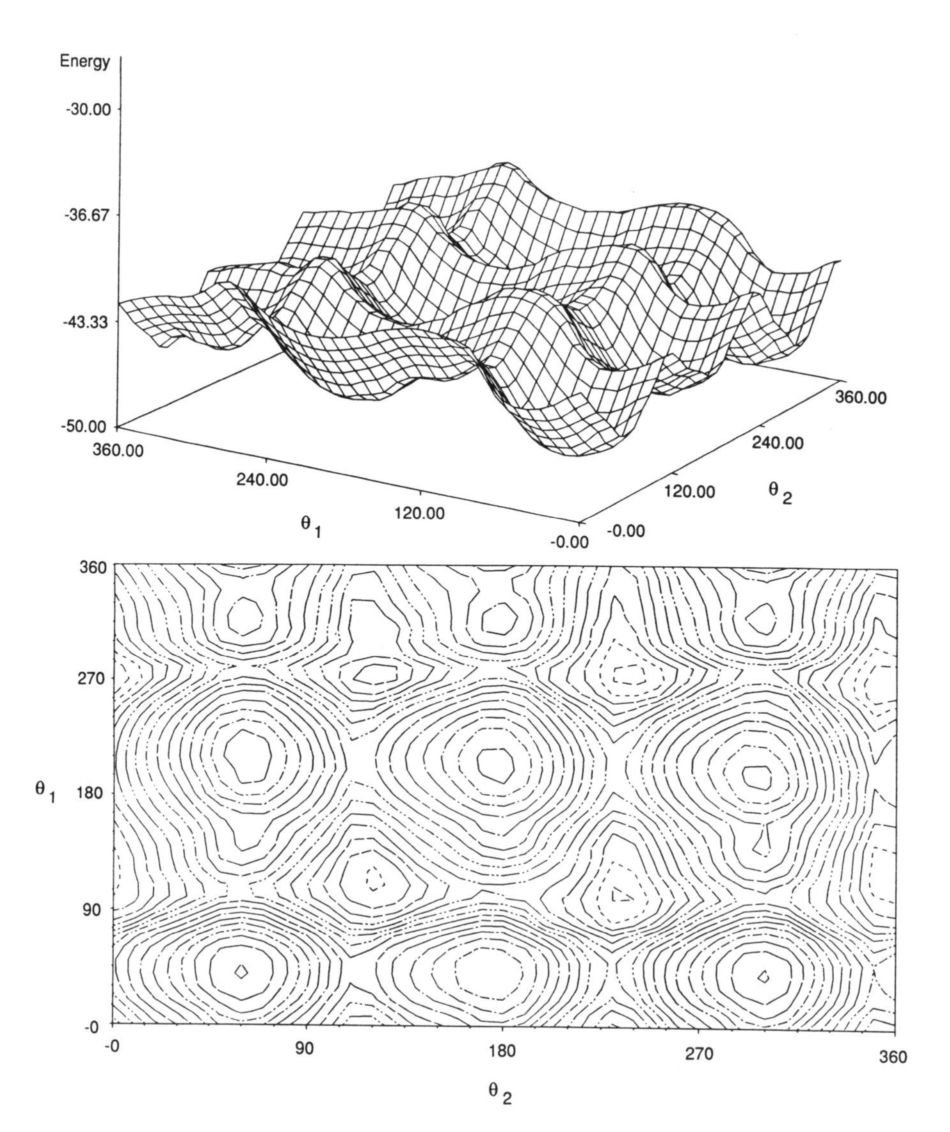

Figure 2. Energy Contour Plot for Thymidine (X-ray Structure) for Two Torsions, θ1 (C8-N9-C1′-O) and θ2 (O-C4′-C5′-O5′)

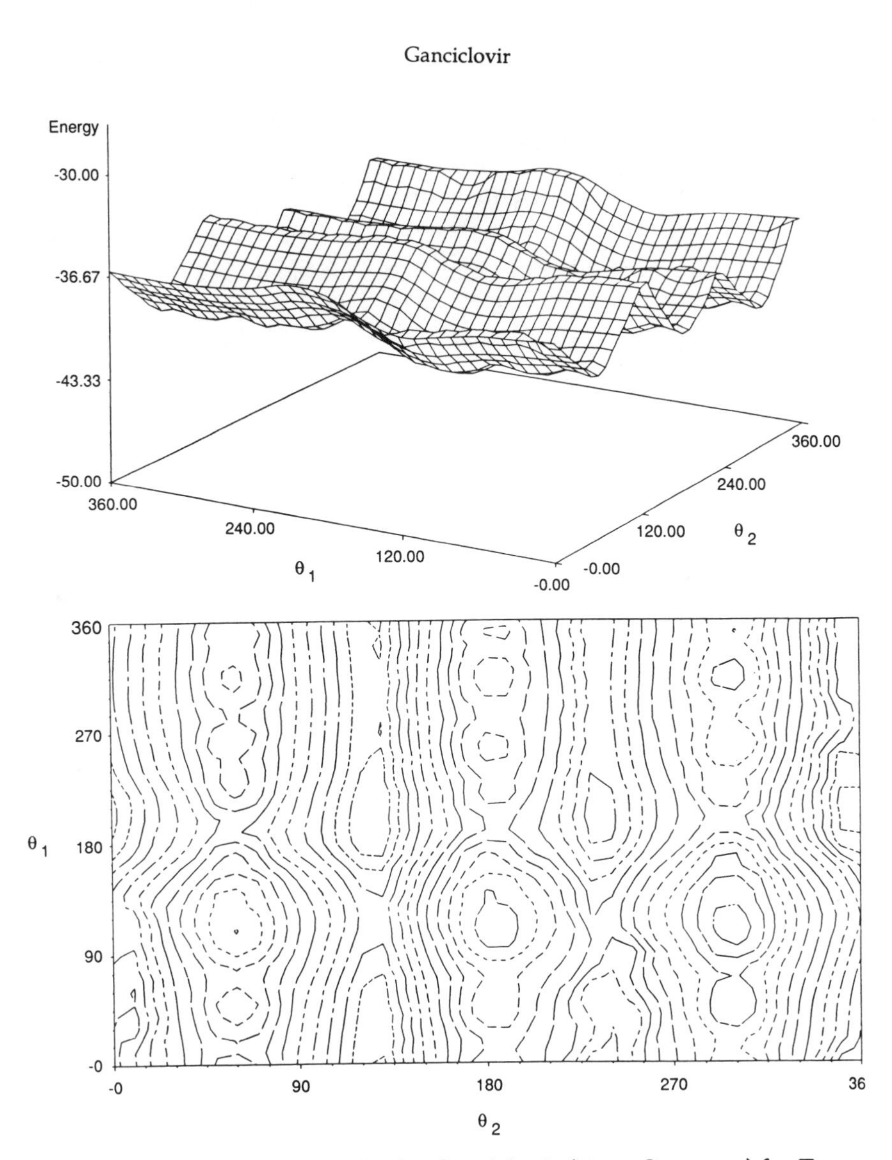

Figure 3. Energy Contour Plot for Ganciclovir (X-ray Structure) for Two Torsions, θ1 (C8-N9-C1′-O) and θ2 (O2′-C3′-C4′-O5′)

deoxycytidine might have predicted the 6-amino derivative to be a substrate].

Similarly, changes were not tolerated at the purine-2-position of guanine, except for the 2-methylamino(6) which retained weak substrate activity. However, hydrophobic substituents at the 8-position (particularly halogen or alkyl) enhanced substrate activity, while hydrophilic substituents were not substrates.

A likely explanation for the substrate activity of 8-substituted-guanines is that the hydrophobic 8-bromo or -methyl fits into the postulated hydrophobic pocket which exists in our model to account for the enhanced substrate activity of 5-substituted thymidines.

Table I. Structure-Substrate Relationship for Modified Guanines

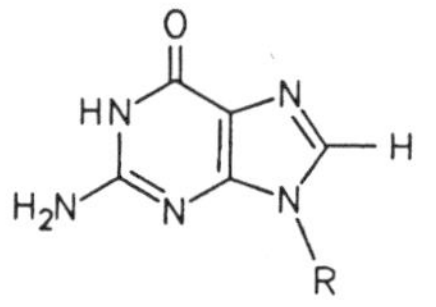

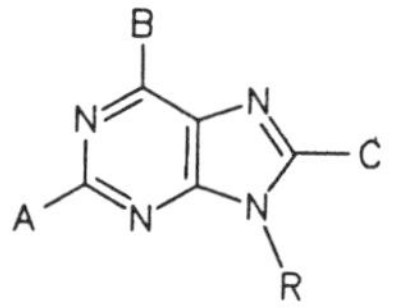

R	A, B, C
R = $CH_2OCH_2CH_2OH$ (+)	A = OH(-), H(-), NHMe(+) $NHCH_2Ph$(-), $NHNH_2$(-), Cl(-)
R = $CH_2OCH(CH_2OH)_2$ (+++)	B = H, SH, NH_2, NHMe, OMe, NHnPr, $NHCH_2Ph$, **all**(-)
R = $CH_2CH_2CH_2CH_2OH$ (++)	C = OH(-), NH_2(-), SH(-) 8-aza(+), Br(++), Me(+++)

NOTE: –, not a substrate; +, poor substrate; ++, good substrate; and +++, excellent substrate. Assay protocol, ref. 21.

In order to test the hypothesis by superposition of molecular models, randomly generated ganciclovir conformational structures were optimized and grouped into families when atoms differed in position by <0.1 Å. The lowest energy family (A141, -42 kcal) when rigidly superimposed upon the X-ray structure of thymidine produced 'scenario A' (Fig. 4), which could account for many of the observed substrate requirements. Five probable points of interaction between the substrate (thymidine) and enzyme were postulated based on this scenario: (a) the hydrophobic pocket at C5, (b) the 3′- and 5′-hydroxyls, (c) the furanose ring oxygen, and (d) the 4-oxo group. In each case the low energy ganciclovir conformer possessed a binding counterpart. Distances between corresponding binding functions of the two molecules in the superposition were generally small (<1.0 Å). The largest displacements of these binding counterparts were (a) the pyrimidine-4-oxo//the guanine-6-oxo (this disparity could perhaps be accommodated if the interacting group on the enzyme were the terminal carboxyl of Asp or Glu), and (b) the 3′-hydroxyl//pro-R hydroxyl (rotation of the GCV pro-*R* hydroxymethyl 60° produced a virtual overlap with the thymidine counterpart). In this binding scenario, the two heteroaromatic systems are nearly coplanar. Moving the purine out of this plane does not markedly effect any of the hypothetical binding interactions except the positioning of the purine C8 substituent into the thymidine C5 hydrophobic pocket.

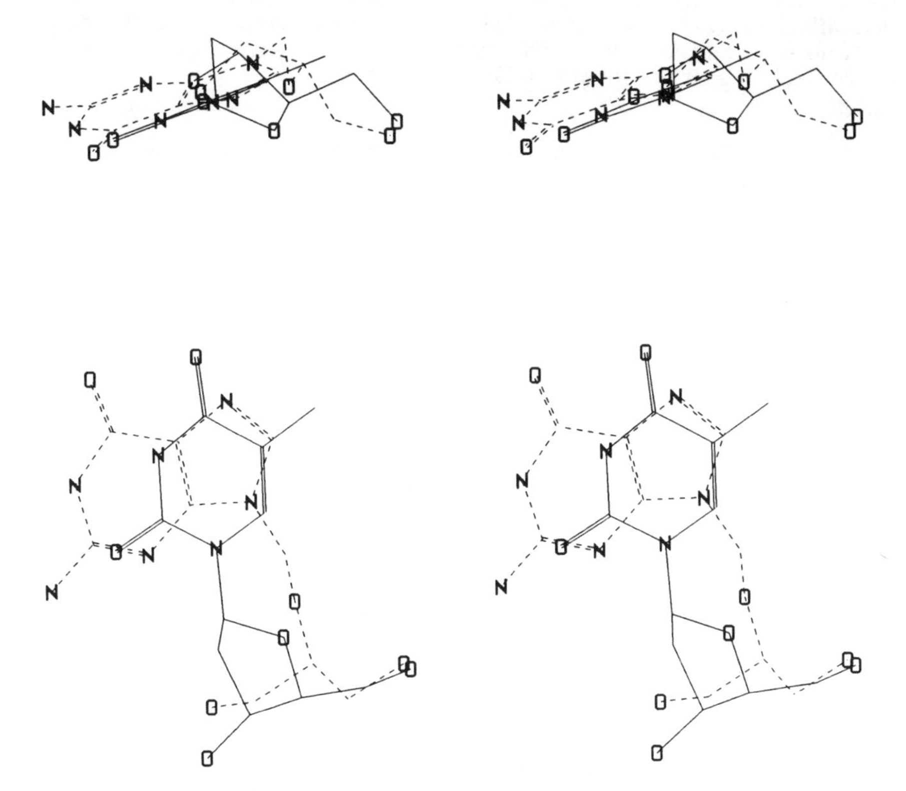

Figure 4. Binding Scenario A: Rigid Superposition of a Low-energy Ganciclovir Conformation (dotted) and Thymidine X-ray Structure (solid)--Front and Top Stereo-Views

Taking note that the N2-3′O distance is 4.5 Å in the superposition suggests that, if that space were a cavity in the enzyme, an N2-substituent bearing a hydrophilic terminus and reaching into the postulated 3′-hydroxyl binding site might pick up substrate activity.

The extended length of the 2-hydroxyethyl group is close to 4.5 Å whereas the 2,3-dihydroxypropyl derivative, which was also prepared, is somewhat longer and may not have the proper geometry for good binding. To check that the N2-hydroxyalkyl was not being phosphorylated by HSV-TK, the two methylated side-chains were prepared. Table II shows that N2-(2-methoxy)ethyl was still a substrate while the 9-(2-methoxy)ethoxymethyl-derivative was not a substrate, lending additional credence to 'binding scenario A′ as a predictive model. These compounds were shown not to be converted to higher phosphates in HSV-infected cells.

Table II. N2-Hydroxyalkyl-guanines as HSV-TK Substrates

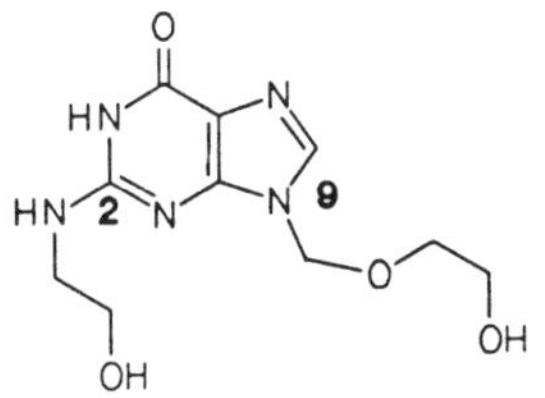

Purine 2-substituent	9-Substituent	Substrate
NH_2	$CH_2OCH_2CH_2OH$	+(ACV)
$NHCH_2CH_2OH$	$CH_2OCH_2CH_2OH$	+++
$NHCH_2CHOHCH_2OH$	$CH_2OCH_2CH_2OH$	+
$NHCH_2CH_2OCH_3$	$CH_2OCH_2CH_2OH$	++
$NHCH_2CH_2OH$	$CH_2OCH_2CH_2OCH_3$	-

NOTE: −, not a substrate; +, poor substrate; ++, good substrate; and +++, excellent substrate. Assay protocol, ref. 21.

Other superpositions did not explain the data as well, but 'binding scenario B′ (Fig. 5) received serious consideration because it was able to accommodate the fact that 2′-deoxycytidine was a good substrate for the enzyme. In this superposition there is excellent overlap of the 2-amino of GCV with the 4-oxo of thymidine as well as good positional agreement between the 3′ and 5′ hydroxyls of each of the two structures. Although the 'fit′ of the two structures is not as good and this is a higher energy conformer of GCV, this superposition clearly predicts that 8-hydroxyalkyl guanines should be substrates. However, when 8-(3-hydroxypropyl)-9-methylguanine(22) and 8-(2-aminoethylamino)-GCV(23) were prepared, they were found not to be substrates although molecular models of these structures were well accommodated in 'binding scenario B′. This scenario also could not account for the fact that an 8-(methyl or bromo) substituent enhanced substrate activity.

Acyclosugar Structure Substrate Activity. A great number of 9-side-chain variants were made by us and others(23-27). It is obvious from Figure 6 that the best substrate activity is achieved when the potential substrate has a

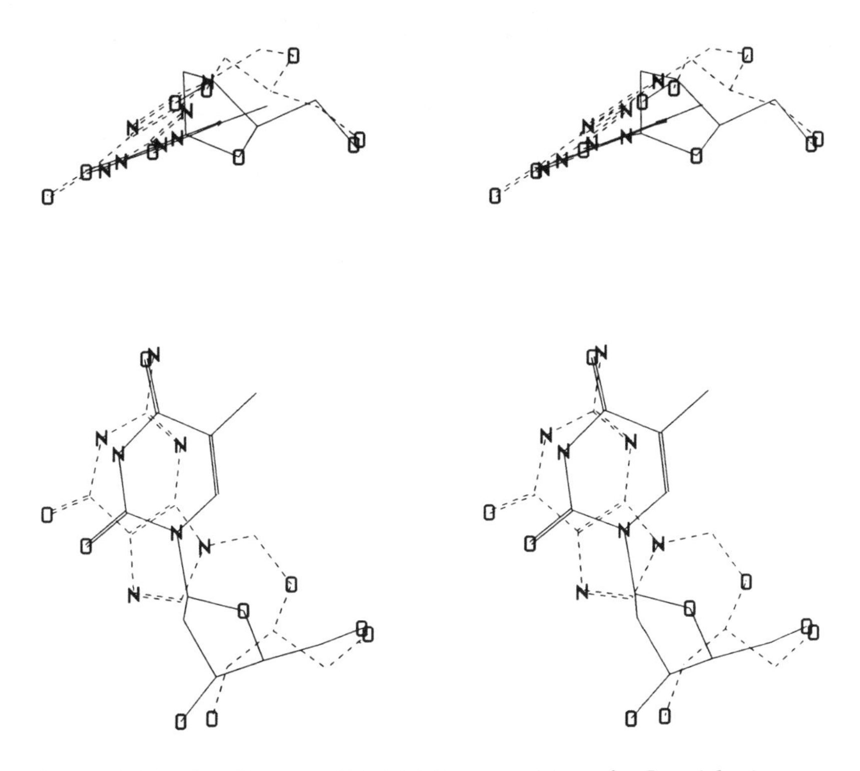

Figure 5. Binding Scenario B: Rigid Superposition of a Ganciclovir Conformation (dotted) and Thymidine X-ray Structure (solid)--Front and Top Stereo-Views

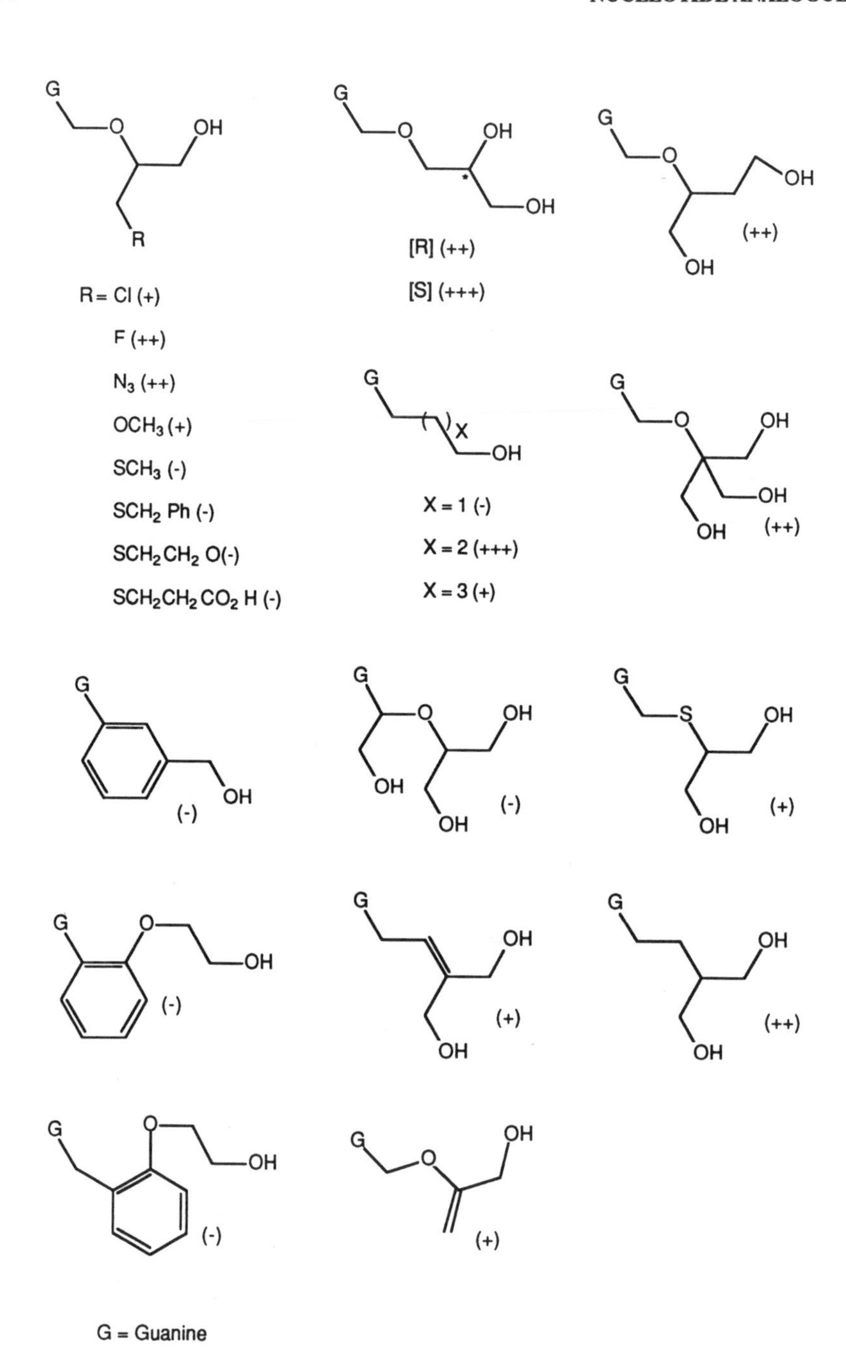

Figure 6. Structure Substrate Relationship for the 9-Side-chain of Guanine Acyclonucleosides

binding functionality in both the 5′ and 3′ hydroxy binding sites for thymidine. Modified GCV's lose significant substrate activity when one of the hydroxyl's is replaced with a large or less hydrophilic group. Substitution of the 1′-position is not tolerated for substrate activity(6) and surprisingly the 2′-carba derivatives are better substrates than their oxa counterparts. The minimum side-chain length for substrate activity is 5 atoms (excluding hydrogen), the ACV side-chain length; longer side-chains are even better substrates for HSV-TK, but are more poorly converted to higher phosphates. There appears to be a chiral preference of the enzyme for the [*S*] configuration at the 4′-position(26) and considerable bulk tolerance. It has been shown that there is poor bulk tolerance on the carbon bearing the hydroxyl which is phosphorylated(26). Some conformationally-constrained acyclonucleosides(28) have been shown to have enhanced substrate activity, but the conformationally-restricted derivatives in which a phenyl ring(23) is employed to give directionality to the side chain do not serve as substrates, possibly due to their excessive bulk. It should be noted that since 'substrate ability' is a measure not only of binding to the active site, but also efficacy on this receptor, that these poor substrates may bind well to the enzyme but lack the proper geometry for efficient phosphorylation.

GMP Kinase

This enzyme which converts the HSV-TK product (monophosphate) to diphosphate(DP) has itself been shown to be very selective(5) in choice of substrates. The chiral GCV-MP's have both been shown to be substrates for GMP kinase but at vastly different rates(26). The [*S*] enantiomer, the one most analogous to the natural configuration, is the better substrate by two orders of magnitude. Figure 7 shows the substrate ability of various monophosphate derivatives. As in the case of thymidine kinase, branching of the acyclo side-chain at 1′ abolished any substrate activity. Unlike TK, which preferred longer side-chains as substrates (8 > 7 > 6 > 5), GMP kinase phosphorylates the shorter side-chains best and 8-membered side-chains not at all. Therefore, the monophosphates of the best TK substrates (with respect to side-chain length) are not substrates for GMP kinase.

The 2′-oxa substitution increases K_m and retards V_{max}. It is not surprising that [*S*] chirality at position 4′ favors phosphorylation by the enzyme since this would aid in the recognition of the natural substrate. A hydroxyl in the region of the 3′-hydroxyl of GMP enhances substrate activity, but phosphorylation of the hydroxyl at this position (3′) abolishes substrate activity.

Conversion to Triphosphate

Use of the 'staggered assay'(21) permits the comparison of the relative phosphorylation rates of various acyclonucleosides. Although all compounds shown in Table III are converted to triphosphate, they differ in the facility with which this occurs. It should also be noted is that not all of the triphosphates (TP) inhibit HSV DNA polymerase and have antiherpes activity. In particular, the unsaturated acyclonucleoside in line 3 is relatively poorly converted to a TP which does not inhibit the viral DNA polymerase, but the compound nonetheless has antiherpes activity (the acyclonucleoside may be an SAH hydrolase inhibitor). It is also interesting that the tris-hydroxymethyl acyclonucleoside and the cyclopropyl compound (lines 4 & 5, respectively) are equally well converted to TP, are also equally devoid of viral polymerase inhibition, and yet only one of the materials (the latter one) still has antiherpes activity by some as yet unidentified mechanism.

Compound	K_m (μM)	V_{max} (nmol/min)
5'-GMP	43	87.5
5'-dGMP	208	36.5
5'-ddGMP	304	1.5
3'-GMP	No Phosphorylation	
G–O–CH2CH2–$OPO_3^=$	625	0.7
G–(CH2)4–$OPO_3^=$	292	0.8
G–O–(CH2)3–$OPO_3^=$	No Phosphorylation	
[S]	537	1.6
[R]	455	1.0
[S,S]	255	0.7

Compound	K_m (μM)	V_{max} (nmol/min)
[S]	58	33.2
[R]	347	3.2
[S]	48	18.8
[R]	214	1.3
	140	20.0
	No Phosphorylation	
[R,R]	No Phosphorylation	
[R,S]	No Phosphorylation	

Figure 7. Kinetic Data for Phosphorylation by GMP Kinase

It is more desirable for an acyclonucleoside to be poorly converted to a triphosphate which is an efficient inhibitor of viral replication (*e.g.* ACV), than one which is efficiently converted to triphosphate. Efficient phosphorylation by viral-specified kinases will undoubtedly mean more phosphorylation (although orders of magnitude less) by host cell kinases in an uninfected cell. Any significant level of phosphorylated acyclonucleoside in uninfected cells can disturb the balance of nucleotide pool sizes(30) and lead to side effects in one of the myriad of various cell types and their metabolic systems for regulation of nucleic acid synthesis.

Table III. Relative Phosphorylation Rates of Acyclonucleosides by HSV-TK and HSV-Infected Cell Extracts and HSV DNA Polymerase Inhibition

	HSV-TK %MP	% MP	DP	TP	% Inh'n HSV DNApol	ED_{50} (µg/ml) HSV-1(29)
ACV	17	3	9	60	79	3
GCV	90	3	10	64	22	0.16
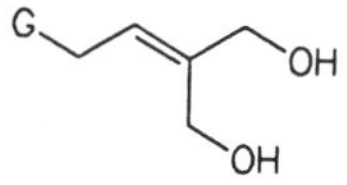 G, OH, OH	8	0	9	25	0	50
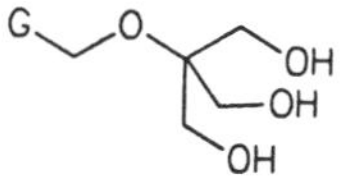 G, O, OH, OH, OH	51	0	8	86	0	>100
 G, OH	45	0	0	76	3	6
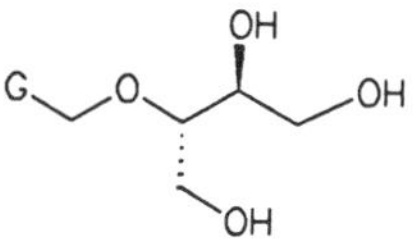 G, O, OH, OH, OH	80	0	5	82	4	>100

2'Nor-cGMP

Still another type of antiherpetic mode of action employing phosphorylated acyclonucleosides is typified by 2'nor-cGMP(31–33). This antiviral agent has potent activity against members of the herpes group and other DNA viruses, including cytomegalovirus(34,35), adenovirus(36), papilloma virus(31), varicella-zoster(31), and herpes keratitis(37). Cell culture studies and animal studies have shown that 2'nor-cGMP is equally effective against TK-deficient (or minus) HSV strains, indicating that this activity in independent of viral thymidine kinase. Although some hydrolysis of the cyclic phosphate occurs intracellularly *in vivo* to produce levels of GCV-MP (and therefore GCV-TP), the amount of TP produced is not enough to account for the antiviral activity observed(38).

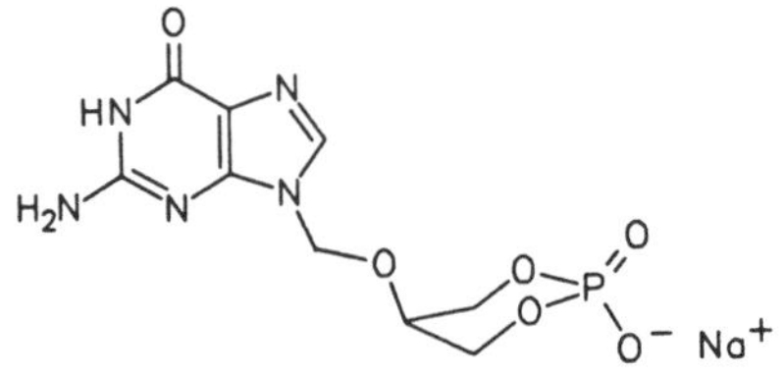

Unlike GCV-MP, 2'nor-cGMP is taken up intact from the gut [10% of a labelled dose (10mg/kg p.o., rats) is recovered in the urine(39)] and is relatively stable in plasma (80% of drug recovered from urine is intact cyclic phosphate).

The mechanism of antiherpetic action of 2'nor-cGMP remains unknown, since the cyclic phosphate directly inhibits neither HSV-1 DNA polymerase nor CMV polymerase.

Summary

Molecular modelling has proven to be a positive factor in the development of structure-activity relationships as a source of stimulation and predictive models. Although the means is available to design acyclonucleoside substrates which will be phosphorylated in herpes-infected cells, there is no assurance that the resultant monophosphates will be converted to higher phosphates by GMP and other kinases. It is also not possible to predict *a priori* whether these triphosphates will inhibit viral polymerases and/or have antiherpetic activity. Other acyclonucleoside or acyclonucleotides have antiviral activity which is independent of DNA polymerase or viral activation by viral thymidine kinase or both.

Acknowledgments

The author is very grateful to the many contributors to this work, especially to Dr. Malcolm MacCoss, Dr. Wallace T. Ashton, and Dr. John D. Karkas who helped to shape and guide this work and to Dr. Dennis Underwood who helped to solve the molecular modelling problems.

Literature Cited

1. Elion, G. B. *Am. J. Med.* **1982**, *73*, 7 (Acyclovir Symposium).
2. Elion, G. B.; Furman, P. A.; Fyfe, J. A.; de Miranda, P.; Beauchamp, L.; Schaeffer, H. J. *Proc. Natl. Acad. Sci. USA* **1977**, *74*, 5716.
3. Fyfe, J. A.; Keller, P. M.; Furman, P. A.; Miller, R. L.; Elion, G. B. *J. Biol. Chem.* **1978**, *253*, 8721.
4. Miller, W. H.; Miller, R. L. *J. Biol. Chem.* **1980**, *255*, 7204.
5. Miller, W. H.; Miller, R. L. *Biochem. Pharmacol.* **1982**, *31*, 3879.
6. Keller, P. M.; Fyfe, J. A.; Beauchamp, L.; Lubbers, C. M.; Furman, P. A.; Schaeffer, H. J.; Elion, G. B. *Biochem. Pharmacol.* **1981**, *30*, 3071.
7. Schildkraut, I.; Cooper, G. M.; Greer, S. *Mol. Pharmacol.* **1975** *11*, 153.
8. Cheng, Y.-C.; Dutschman, G.; De Clercq, E.; Jones, A. S.; Rahim, S. G.; Verhelst, G.; Walker, R. T. *Mol. Pharmacol.* **1981**, *20*, 230.
9. Sim, I. S.; Goodchild, J.; Meredith, D. M.; Porter, R. A.; Raper, R. H.; Viney, J.; Wadsworth, H. J. *Antimicrob. Ag. Chemother.* **1983**, *23*, 416.
10. Cheng, Y.-C. *Ann. N. Y. Acad. Sci.* **1977**, *284*, 594.

11. Cheng, Y.-C.; Ostrander, M. *J. Biol. Chem.* **1976**, *251*, 2605.
12. Karkas, J. D.; Germershausen, J. G.; Tolman, R. L.; M. MacCoss; Wagner, A. F.; Liou, R.; Bostedor, R. *Biochem. Biophys. Acta* **1987**, *911*, 127.
13. Marshall, G.; *Trends Biol. Sci.* **1988**, 236.
14. The MM2-extended force field (MM2-X), developed internally at Merck, shares many parameters with MM2; it differs principally in that lone pairs on heteroatoms are not used (compensating changes have been made in torsional parameters and charge distributions), and that electrostatic interactions take place between atom-centered charges, thus allowing proper treatment of charged systems. MM2-X has been parameterized for a wide range of systems.
15. Birnbaum, G. I.; Shugar, D. in Nucleic Acid Structure, Part 3, Topics in Molecular and Structural Biology, ed. S. Neidle, MacMillan (London, 1987) p. 1.
16. Birnbaum, G. I.; Cygler, M.; Kusmierek, J. T.; Shugar, D. *Biochem. Biophys. Res. Commun.* **1981**, *103*, 968.
17. Birnbaum, G. I.; Cygler, M.; Shugar, D. *Can. J. Chem.* **1984**, *62*, 2646.
18. Birnbaum, G. I.; Johansson, N. G.; Shugar, D. *Nucleosides & Nucleotides* **1987**, *6(4)*, 775.
19. Stolarski, R.; Lassota, P.; Kazimierczuk, Z.; Shugar, D. Z. *Naturforsch* **1988**, *43c*, 231.
20. Beauchamp, L. M.; Dolmatch, B. L.; Schaeffer, H. J.; Collins, P.; Bauer, D. J.; Keller, P. M.; Fyfe, J. A. *J. Med. Chem.* **1985**, *28*, 982.
21. Ashton, W. T.; Canning, L. F.; Reynolds, G. F.; Tolman, R. L.; Karkas, J. D.; Liou, R.; Davies, M.-E. M.; DeWitt, C. M.; Perry, H. C.; Field, A. K. *J. Med. Chem.* **1985**, *28*, 926.
22. Stein, J. M.; Stoeckler, J. D.; Li, S.-Y.; Tolman, R. L.; MacCoss, M.; Chen, A.; Karkas, J. D.; Ashton, W. T.; and Parks, R. E., Jr. *Biochem. Pharmacol.* **1987**, *36*, 1237.
23. Tolman, R. L.; Ashton, W. T.; MacCoss, M.; Karkas, J. D.; Underwood, D.; Meurer, L. C.; Cantone, C. G.; Johnston, D. B. R.; Hannah, J.; Liou, R. *J. Med. Chem.* 1989, submitted.
24. Martin, J. C.; McGee, D. P.C.; Jeffrey, G. A.; Hobbs, D. W.; Smee, D. F.; Matthews, T. R.; Verheyden, J. P. H. *J. Med. Chem.* **1986**, *29*, 1384.
25. McGee, D. P. C.; Martin, J. C.; Smee, D. F.; Matthews, T. R.; Verheyden, J. P. H. *J. Med. Chem.* **1985**, *28*, 1242.
26. Karkas, J. D.; Ashton, W. T.; Canning, L. F.; Liou, R.; Germershausen, J.; Bostedor, R.; Arison, B.; Field, A. K.; Tolman, R. L. *J. Med. Chem.* **1986**, *29*, 842.
27. Larsson, A.; Tao, Pei-zhen *Antimicrob. Ag. Chemother.* **1984**, *25*, 524.
28. Ashton, W. T.; Meurer, L. C.; Cantone, C. L.; Field, A. K.; Hannah, J.; Karkas, J. D.; Liou, R.; Patel, G. F.; Perry, H. C.; Wagner, A. F.; Walton, E.; Tolman, R. L. *J. Med. Chem.* **1988**, *31*, 2304.
29. Field, A. K.; Perry, H. C. personal communication.
30. Bradley, M. O.; Sharkey, N. A. *Nature* **1978**, *274*, 608.
31. Tolman, R. L.; Field, A. K.; Karkas, J. D.; Wagner, A. F.; Germershausen, J.; Crumpacker, C.; Scolnick, E. M. *Biochem. Biophys. Res. Comm.* **1985**, *128*, 1329.
32. Field, A. K.; Davies, M.-E.; DeWitt, C. M.; Perry, H. C.; Schofield, T. L.; Karkas, J. D.; Germershausen, J.; Wagner, A. F.; Cantone, C. L.; MacCoss, M.; Tolman, R. L. *Antiviral Res.* **1986**, *6*, 329.
33. Prisbe, E. J.; Martin, J. C.; Baker, M. F.; Smee, D. F.; Duke, A. E.; Matthews, T. R.; Verheyden, J. P. H. *J. Med. Chem.* **1986**, *29*, 671.
34. Duke, A. E.; Smee, D. F.; Chernow, M.; Boehme, R.; Matthews, T. R. *Antiviral Res.* **1986**, *6*, 299.

35. Yang, Z. H.; Lucia, H. L.; Hsiung, G. D.; Tolman, R. L.; Colonno, R. J. *Antimicrob. Ag. Chemother.* **1989** submitted.
36. Baba, M.; Mori, S.; Shigeta, S.; DeClercq, E. *Antimicrob. Ag. Chemo.* **1987**, *31*, 337.
37. Gordon, Y. J.; Capone, A.; Sheppard, J.; Gordon, A.; Romanowski, E.; Araullo-Cruz, T *Current Eye Res.* **1987**, *6*, 247.
38. Germershausen, J. G.; Liou, R.; Field, A. K.; Wagner, A. F.; MacCoss, M.; Tolman, R. L.; Karkas, J. D. *Antimicrob. Ag. Chemother.* **1986**, *29*, 1025.
39. Hucker, H.; Germershausen, J. personal communication.

RECEIVED June 9, 1989

Chapter 4

Phosphonylmethyl Ethers of Nucleosides and Their Acyclic Analogues

Antonin Holy[1], Erik De Clercq[2], and Ivan Votruba[1]

[1]Institute of Organic Chemistry and Biochemistry, Czechoslovak Academy of Sciences, 166 10 Praha 6, Czechoslovakia
[2]Rega Institute, Katholieke Universiteit Leuven, Minderbroedersstraat 10, B–3000 Leuven, Belgium

Structure-activity investigation in the series of acyclic nucleotide analogs bearing a modified phosphoric acid residue at the side-chain revealed two novel classes of antivirals: N--(3-hydroxy-2-phosphonylmethoxypropyl)- (HPMP-) and N-(2-phosphonylmethoxyethyl)(PME-) derivatives of heterocyclic bases. Adenine, guanine, 2-aminoadenine and (in the HPMP-series) cytosine derivatives act specifically against DNA viruses (herpes viruses,adenoviruses,poxviruses). The PME-compounds are also active against retroviruses (MSV,HIV)and exhibit a cytostatic effect on L-1210 mouse leukemia cells.The drugs are converted by the action of cellular nucleotide kinases into their diphosphates and inhibit viral and, to a lesser extent, cellular DNA synthesis. These metabolites exert a comparatively low inhibitory effect on viral (HSV-1) DNA polymerase. The diphosphates derived from PME-compounds strongly inhibit viral (HSV-1) ribonucleotide reductase and AMV reverse transcriptase.

The majority of biologically active nucleoside analogs exert their effects via their 5'-nucleotides. However, direct application of the latter compounds does not bring about any substantial improvement of the activity except for certain cases in which the nucleotide has more favorable pharmacological parameters (increased solubility, etc.) than the parent nucleoside (1). This apparent paradox is easily explained by rapid enzymatic dephosphorylations which occur in the serum, in the cellular membranes and in the cytoplasm.In order to circumvent these catabolic reactions, many attempts have been undertaken to synthesize analogs of nucleotides with mo-

0097–6156/89/0401–0051$06.25/0

dified phosphoric acid ester linkage which would resist both chemical as well as enzymatic hydrolysis (2). An exhaustive investigation during the past two decades resulted in the design of numerous analogs which were studied in many respects including evaluation of their biological activities as well as inhibition of enzymatic reactions in vitro. Though some of the analogs exhibited interesting inhibitory activities against some isolated enzymes (1), none of them was found to posses significant antiviral, anticancer or other biological effects.

These studies complemented by detailed investigation of structural specificities of selected phosphomonoester hydrolases, uncovered **isopolarity** as one of the major important structural features in this field: In addition to steric similarity, active analogs must always have the same dissociability as their phosphate counterparts. The wide occurence of non-specific dephosphorylation enzymes capable of splitting any phosphoric acid ester bond disregarding the nature of the alcohol component leaves very narrow field for chemical variation of this linkage. The only type of compounds which are analogous to phosphomonoesters [**1**], fulfill the above conditions and completely resist the enzymatic cleavage, are evidently the compounds bearing a phosphoric acid residue linked directly to the sugar residue of a nucleoside (e.g. [**2**])by a C-P linkage.

$RO\text{-}P(O)(OH)_2$ **1** $R\text{-}P(O)(OH)_2$ **2** $R\text{-}OCH_2P(O)(OH)_2$ **3**

(R=5'-deoxynucleoside-5'-yl residue)

PHOSPHONYLMETHYL ETHER DERIVATIVES.

The above mentioned biological inactivity of the phosphonate analogs of nucleotides (type [**2**])as well as our preceding experience on the structural requirements of nucleolytic enzymes (3) which led us to realize the importance of the oxygen atom in the vicinity of phosphorus stimulated us to investigate a novel class of phosphorus-modified nucleotide analogs which are both isopolar and nearly isosteric with their phosphate counterparts. These derivatives (type [**3**]) contain an enzymatically resistent ether linkage joining the hydroxyl of a nucleoside sugar moiety with the phosphoric acid grouping by means of a methylene unit. This unit enables an adaptation of the molecular conformation not substantially changing the pK value of the resulting phosphonylmethyl ether in comparison with the corresponding phosphomonoester.

5'-O-Phosphonylmethyl Nucleosides. The first group of analogs investigated in this series encompasses compounds derived from naturally occuring ribonucleosides by an etherification of their 5'-hydroxyl group. Their synthesis was accomplished by treatment of a nucleoside alkoxide anion (generated by sodium hydride reaction) with dialkyl p-toluenesulfonyloxymethylphosphonate [**4**].

The intermediary diester is then cleaved by bromotrimethylsilane treatment followed by acid hydrolysis and

$$RO^{-} + TsOCH_2P(O)(OAlk)_2 \longrightarrow [ROCH_2P(O)(Alk)_2] \longrightarrow [3]$$

4

(R=ribonucleosid-5'-yl residue)

Scheme 1

removal of the protecting group (Scheme 1) (4). These compounds resist the action of phosphomonoesterases (3); also the internucleotidic linkage containing this grouping (5) as well as the cyclic esters of this type (6) are stable against the relevant ribonucleases and phosphodiesterases. The substrate and inhibitory activity of ribonucleoside 5'-triphosphate analogs containing modified α-phosphorus atom, as observed with uridine kinase (7,8),polynucleotide phosphorylase (9) and DNA-dependent RNA polymerase (10,11), suggest that these compounds might replace the 5'-phosphates of antiviral or anticancer nucleoside analogs at their target enzymes and interfere with metabolic pathways of nucleotides. The determination of antiviral activity against both RNA and DNA viruses disproved this expectation for 5'-O-phosphonylmethyl ribonucleosides (E.De Clercq and A.Holý, unpublished data). We have also prepared the 5'-O-phosphonylmethyl ethers of those sugar-modified nucleoside analogs which *per se* exhibit an outstanding antiviral and/or cytostatic effect: 6-azauridine, cytosine arabinoside, adenine arabinoside, thymine arabinoside, adenine xyloside and Ribavirin (J.Brokeš, A.Holý, J.Zajíček, *Collect.Czech. Chem.Commun.*, in press) by the above method. However, all of them are devoid of antiviral activity *in vitro* against all viruses tested (E.De Clercq: unpublished data). The reason for this failure remains to be elucidated; since all the above compounds exert only marginal (if any) cytotoxicity on the host cell lines, it might be anticipated that they are unable to penetrate the cell membrane to the extent that is required to cause a biological effect.

Phosphonylmethyl Ethers of Acyclic Nucleotides. Acyclic nucleoside analogs in which the carbohydrate moiety is replaced by an open chain mimicking the part of the pentafuranose ring, have attracted a major interest in many laboratories including ours. The development in this field was largely stimulated by the antiviral activity found for numerous representatives of this group. In the guanine series, acyclovir [**5**] and its congeners (ganciclovir [**6**], buciclovir [**7**])are active against herpes viruses, whereas the adenine derivatives 9-(S)-(2,3-dihydroxypropyl)adenine (DHPA) [**8**] and 3-(adenin-9-yl)-2--hydroxypropanoic acid and its esters [**9**] inhibit the multiplication of certain DNA and RNA viruses with a preference for (-)-stranded and (+)-stranded RNA viruses (12-16). The activity of the former group of antivirals, i.e. acyclovir and all of its analogs,depends on their phosphorylation to monophosphates which then undergo further transformations resulting ultimately in the inhibition of viral DNA synthesis. This phosphorylation is catalyzed by a virus-specific "thymidine kinase" (TK) (12). An artificial introduction of the phosphate group or of its (enzymatically stable) structural equivalent might circumvent the specificity of these drugs for TK^+ viruses, albeit, with the loss of high selectivity of action.

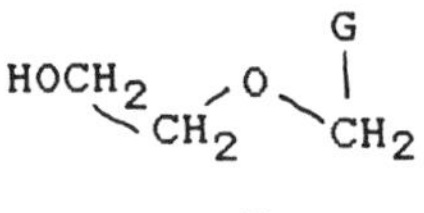

```
                 G                        G                         G
HOCH2  ,O\      |        HOCH2  ,O\      |        HOCH2       ,CH2\  |
    \CH2   \CH2              \CH    \CH2              \CH2/       \CH2
                          HOCH2/
        5                         6                         7
```

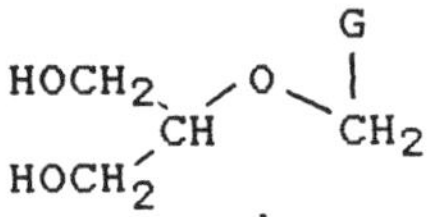

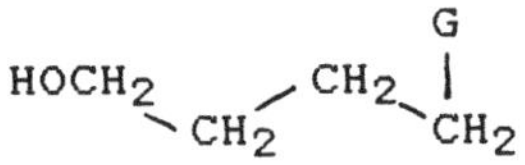

The 1,3-cyclic phosphate [10] derived from ganciclovir [9-(1,3-dihydroxy-2-propoxymethyl)guanine] displays significant inhibitory activity against CMV and other DNA viruses (17). This discovery fully supports the above hypothesis. Also, the marked antiviral activity which has been described for various 9-(ω-phosphonylalkyl)hypoxanthines (18) (the "parent" ω-hydroxyalkyl compounds being inactive) suggests that analogs of nucleotides with modified carbohydrate moiety and phosphate linkage might exert biological activities. These effects also indicate that the polarity of the phosphonate containing molecule might not be the ultimate obstacle of their penetration and action.

Antiviral or other biological activities of acyclic nucleoside analogs derived from adenine do not depend on any preceding phosphorylation (19). These analogs are not phosphorylated either _in vitro_ (20) or _in vivo_ (21) and their phosphates (e.g. from DHPA) do not mediate the activity shown by the parent compound. In this context, we considered it of interest to investigate the O-phosphonylmethyl ethers of DHPA enantiomers as analogs which cannot undergo dephosphorylation. Starting from the racemic compound [8] and utilizing the method described by Scheme 1, which in the case of vicinal diols (22) affords a mixture of the two position isomers, we obtained a racemic mixture of two isomeric O-phosphonylmethyl ethers derived from DHPA. This product exhibited an exceptionally high antiviral activity directed specifically against DNA viruses (23). No effect was observed against any RNA viruses, in particular those with the greatest sensitivity towards DHPA (_vide supra_).

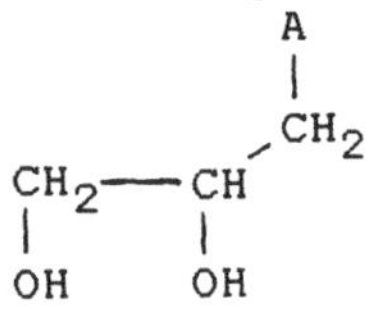

```
            A                   A                              G
            |                   |                              |
            ,CH2               ,CH2          O   ,O-CH2    ,O\    |
CH2----CH/          ROOC-CH/               \\P/      \CH/     \CH2
|       |                 |                 / \      /
OH      OH                OH              HO   O-CH2
      8                   9                       10
```

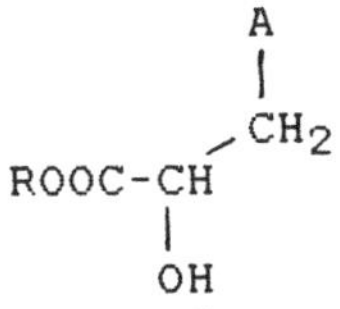

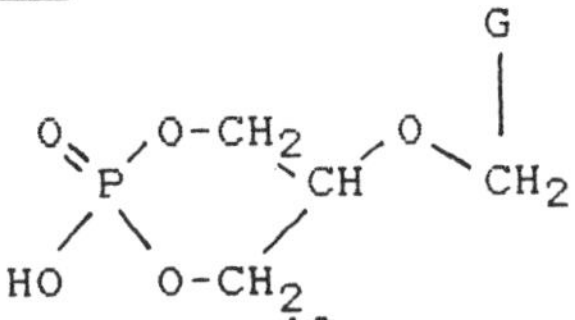

Structure-activity Investigation of Acyclic Nucleotide Analogs.
This discovery prompted us to investigate in detail additional acyclic nucleotide analogs which contain a modified phosphate (phosphonate) group linked to the acyclic side chain by other than ester linkage. The structure-activity investigation was divided into three main groups: (a) identification of isomer(s) responsible for antiviral activity of DHPA phosphonylmethyl ethers, (b) changes of the side-chain which were introduced systematically in the series of 9-substituted adenine derivatives and (c)variation of the heterocyclic base in the series of 2,3-dihydroxypropyl derivatives and any additional series which might arise from the side-chain variation.

Phosphonylmethyl Ethers of 9-(2,3-Dihydroxypropyl)adenine. The synthesis of the four possible isomers of the title compound makes use of a reaction originally developed for the preparation of 2'(3')-O-phosphonylmethyl ribonucleosides (22): the reaction sequence described by Scheme 2 starts from (S)- or (R)-DHPA (24) which on treatment with chloromethylphosphonyl dichloride affords a mixture of chloromethylphosphonates [**11,12**]. These isomers can be separated by HPLC or ion exchange chromatography and further transformed by heating with an aqueous alkali to the phosphonylmethyl ethers [**13,14**]. This transformation which proceeds *via* an intramolecular cyclisation mechanism is not accompanied by any isomerisation or racemization (25):

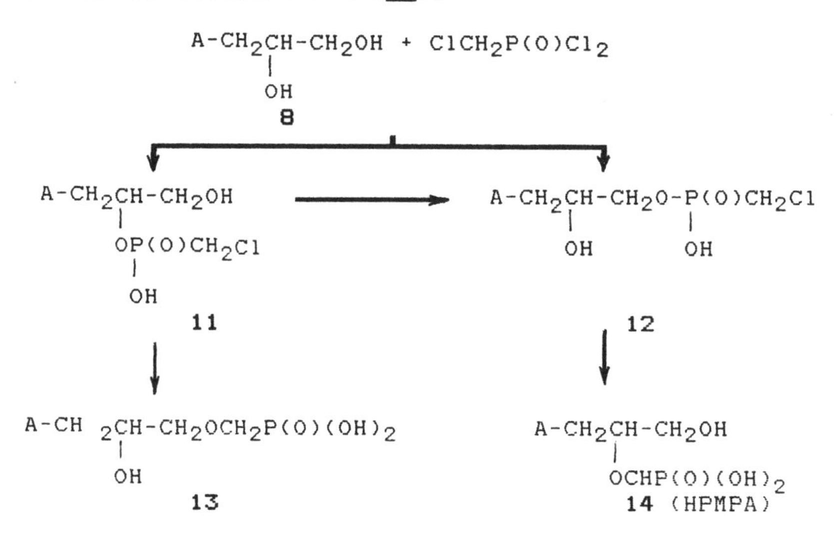

Scheme 2

The evaluation of antiviral activities of the four stereoisomers revealed that the 2'-isomer of the (S)-series [14] i.e.9-(S)-(3-hydroxy-2-phosphonylmethoxypropyl)-adenine is entirely responsible for the antiviral activity, whereas the (S)-3'-isomer as well as the two (R)-enantiomers are devoid of any antiviral efficiency (23). The compound (S)-[14] is currently abbreviated as **HPMPA**.

Variation of the Side Chain. The chemical modifications of the side-chain of the parent structure [14]concerned several of its important structural features. For the sake of simplicity all compounds were derived from adenine which was substituted at the position N9. Chemical syntheses consisted of two general approaches: introduction of a phosphorus-containing grouping into corresponding 9-alkyladenine backbone (using suitable protection strategy) or, alkylation of adenine by a preformed phosphorus-containing synthon bearing a suitable leaving group (alkyl halogenide or alkyl p-toluenesulfonate type synthons). These syntheses are described in the original paper [26]. This series is composed of compounds which bear the phosphonylmethyl ether residue (Table 1) as well as of additional phosphonylalkoxyalkyl- and phosphonylalkyl derivatives lacking the oxygen atom in the vicinity of the phosphorus (Table 2).

Evaluation of the antiviral activity of these compounds (vide infra) revealed that an etherification of the primary hydroxyl group in HPMPA(compounds [15,16]) or its replacement by methyl group (as in [17])resulted in the loss of activity. Also the relative distance of the phosphonylmethoxy group from the base is of ultimate importance,as witnessed not only by the lack of activity of the 3'-isomer of HPMPA [13], its deoxy analog [20], methyl ether [27], homologs [28-30] and 1'-C-alkyl derivatives [32,33] but most evidently by the homolog of HPMPA in which the whole grouping with the phosphonate part of the side chain is shifted by a methylene unit apart from the base [26]. Also, any substitution at the 2'- or 3'-position of the HPMPA skeleton [23-25] resulted invariably in a complete loss of antiviral activity.

Though the above interpretation stresses the absolute importance of the hydroxymethyl group in HPMPA, the study of homologous 9-(ω-phosphonylmethoxyalkyl)adenines unveiled the antiviral activity of 9-(2-phosphonylmethoxyethyl)adenine [19], which is in many respects qualitatively and quantitatively comparable with that of HPMPA (23). This compound will be referred to as **PMEA** and its base modified congeners as PME-derivatives. PMEA can be regarded as a simplified analog of HPMPA lacking the hydroxymethyl group.Neither the phosphonylmethoxymethyl- [18] nor other straight-chain w-phosphonylmethoxyalkyl (propyl, butyl, pentyl) derivatives of adenine [20-22] exhibit any activity whatsoever, and substitution of the two-carbon atom chain by hydroxymethyl group, which in position 2 of the chain leads to HPMPA, is totally ineffective in the position 1 (compound [31]).

However, the presence of α-phosphonyl group and its mutual orientation with respect to the adenine base is not sufficient to cause antiviral activity in the above compounds. The isomer of PMEA, 2-phosphonylethoxymethyl derivative [34] and the carba analog, 9-(4-phosphonylbutyl)adenine [38] exert no antiviral activity. Nor was

Table 1. 9-(Phosphonylmethoxyalkyl)adenines

No.	Side Chain
15	$-CH_2CH(OCH_2P(O)(OH)_2)CH_2OCH_3$
16	$-CH_2CH(OCH_2P(O)(OH)_2)CH_2OC_8H_{17}$
17	$-CH_2CH(OCH_2P(O)(OH)_2)CH_3$
18	$-CH_2OCH_2P(O)(OH)_2$
19	$-CH_2CH_2OCH_2P(O)(OH)_2$
20	$-CH_2CH_2CH_2OCH_2P(O)(OH)_2$
21	$-CH_2(CH_2)_2CH_2OCH_2P(O)(OH)_2$
22	$-CH_2(CH_2)_3CH_2OCH_2P(O)(OH)_2$
23	$-CH_2CH(OCH_2P(O)(OH)_2)CH(OH)CH_2OH$
24	$-CH_2C(CH_3)(OCH_2P(O)(OH)_2)CH_2OH$
25	$-CH_2CH(OCH_2P(O)(OH)_2)CH(OH)CH_3$
26	$-CH_2CH_2CH(OCH_2P(O)(OH)_2)CH_2OH$
27	$-CH_2CH(OCH_3)CH_2OCH_2P(O)(OH)_2$
28	$-CH_2CH_2CH(OH)CH_2OCH_2P(O)(OH)_2$
29	$-CH_2CH(OH)CH(OH)CH_2OCH_2P(O)(OH)_2$
30	$-CH_2CH(OH)CH(CH_3)OCH_2P(O)(OH)_2$
31	$-CH(CH_2OH)CH_2OCH_2P(O)(OH)_2$
32	$-CH(CH_3)CH(OH)CH_2OCH_2P(O)(OH)_2$
33	$-CH(C_6H_{11})CH(OH)CH_2OCH_2P(O)(OH)_2$
34	$-CH_2OCH_2CH_2OCH_2P(O)(OH)_2$

there any antiviral activity observed with the simplified molecules [36,37] or with their homologs with or without an ether linkage in the chain [34,38]. Thus, it can be concluded that the oxygen atom of the phosphonylmethoxy group plays a determining role in the antiviral activity of the acyclic nucleotide analogs. The sole presence of the oxygen atom in the vicinity of phosphorus is not sufficient, as witnessed by the inactivity of compounds [40] and [41].

The modification at the side-chain defined a narrow structural margin in the design of acyclic nucleotide analogs and limited our subsequent studies to the investigation of base-modified analogs of HPMPA and PMEA.

<u>Base-modified N-(3-Hydroxy-2-phosphonylmethoxypropyl) derivatives of heterocyclic bases (HPMP-derivatives)</u>. In order to investigate the effect of the heterocyclic base on the antiviral activity, purine and pyrimidine derivatives of the type [14] were synthesized at first from the racemic N-(2,3-dihydroxypropyl) derivatives

Table 2. 9-(Phosphonylalkyl)adenines

No.	Side Chain
35	$-CH_2OCH_2CH_2P(O)(OH)_2$
36	$-CH_2P(O)(OH)_2$
37	$-CH_2CH_2P(O)(OH)_2$
38	$-CH_2(CH_2)_2P(O)(OH)_2$
39	$-CH_2CH(OH)CH(OH)CH_2P(O)(OH)_2$
40	$-CH_2CH(OH)P(O)(OH)_2$
41	$-CH_2CH_2CH(OH)P(O)(OH)_2$

easily accessible by the alkylation of heterocyclic bases (24). The method used followed exactly that described by Scheme 2. In those cases where the preliminary investigation confirmed a significant biological activity, the pure 2'-isomers were prepared in the (S)-series by a route which affords unequivocally the single isomer via specifically protected intermediates (Scheme 3):

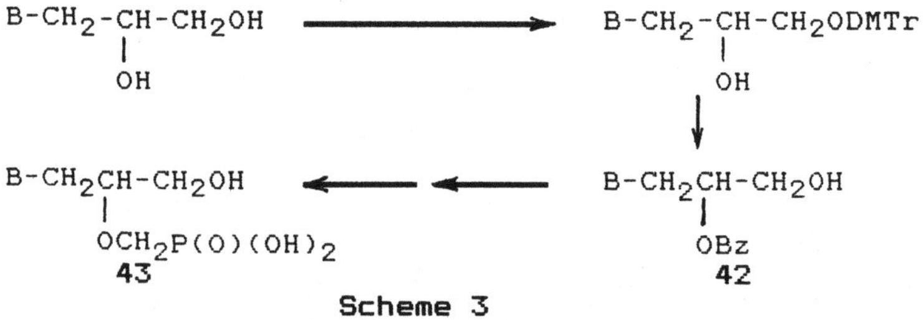

Scheme 3
(Bz...benzoyl, DMTr...dimethoxytrityl group)

This method developed originally for HPMPA (25) makes use of a specific protection by the dimethoxytrityl group at the primary hydroxyl of a base-protected (benzoylated) starting material. The subsequent benzoylation and detritylation afford the 2-O-benzoate [**42**] which on successive treatment with chloromethylphosphonyl dichloride followed by reaction with aqueous alkali affords (under simultaneous debenzoylation) the required compound [**43**]. This reaction was successfully applied for the preparation of adenine, 2-aminopurine, 2,6-diaminopurine, guanine and other derivatives, whereas in the case of cytosine a modification of this procedure was required to eliminate the alkaline treatment which would cause deamination to the uracil derivative (Holý, A., Rosenberg, I., Dvořáková, H.: Collect.Czech.Chem.Commun. in press). The compounds [**43**] prepared by the above pro cedures are listed in the Table 3.

Another strategy for preparation of HPMPA was described in the literature (27): this route starts from N^6,O^3-ditrityl derivative of (S)-DHPA which reacts with the phosphorus synthon [**4**] according to the Scheme 1 and affords HPMPA on subsequent treatment of the protected intermediate with bromotrimethylsilane and aqueous acid Similar method makes use of N^6-benzoyl-3'-O-trityl-(S)--DHPA as a starting material (Holý, A., Rosenberg, I., Dvořáková, H.: Collect.Czech.Chem.Commun. in press).

An alternative general procedure for the synthesis of HPMP-derivatives is based on alkylation of the heterocyclic bases with chiral synthons which possess a pre-formed structure of the side-chain. Such a method was used for the synthesis of HPMPC; the key-compound, tosylate (or mesylate) of diethyl (2S)-(3-benzyloxy-1-hydroxy-2-propoxy)methylphosphonate was prepared from (R)--glycerol acetonide and synthon [**4**] by a multi-step procedure (28).

N-(2-Phosphonylmethoxyethyl) Derivatives of Heterocyclic Bases (PME-Derivatives). The original procedure for the preparation of compound [**19**] (PMEA) which consisted in the treatment of 9-(2-hydroxyethyl)adenine with the synthon [**4**] and subsequent hydrolysis (29) was replaced by an alternative procedure represented in Scheme 4.

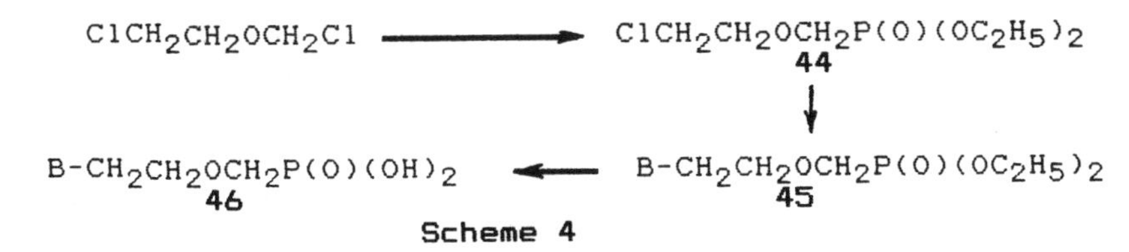

Scheme 4

The chloromethylation of 2-chloroethanol by 1,3,5-trioxane/HCl affords an intermediate which in an Arbuzov reaction with triethyl phosphite unequivocally gives the stable phosphorus-synthon [**44**] (this synthesis is more favorable than the alternative route starting from diethyl 2-hydroxyethoxymethylphosphonate (29)). Alkylation of the heterocylic base (preferably its sodium salt) by this reagent affords compound [**45**] which is easily converted to the final product [**46**]by bromotrimethylsilane treatment (Holý,A., Rosenberg,I., Dvořáková,H.: Collect. Czech.Chem.Commun., in press). The base-modified PME-derivatives [**46**]which were obtained by this procedure are listed in Table 3.

Antiviral activity.

HPMPA is effective against all major herpesviruses at concentrations which are far below the toxicity for the host cells (23). Its potent activity was demonstrated not only with HSV-1 and HSV-2; also animal herpes viruses, e.g.suid, bovid and equid herpesviruses type 1 (23) as well as seal herpes virus (30) are sensitive towards the action of this analog. Epstein-Barr virus (31) and all strains of varicella-zoster viruses and cytomegaloviruses tested so far were also effectively inhibited (23,32,33).Most importantly,HPMPA is equally active against wild-type(TK^+)and TK^- mutant strains of herpes

viruses. In that respect, HPMPA and its congeners markedly differ from the majority of anti-HSV agents the activity of which depends on phosphorylation by (virus-encoded) thymidine kinase.
Poxviruses (vaccinia virus, 23), iridoviruses (African swine fever virus, 34) and adenoviruses (35) also respond very markedly to HPMPA treatment. In some cases, e.g. with VZV (32) or ASFV (34) a selectivity index of 4 orders of magnitude was achieved.

Table 3. Base-modified HPMP- and PME-derivatives of the general formula $B\text{-}CH_2CH(R)OCH_2P(O)(OH)_2$

B [a]	[43]	Abbrev. (R=CH_2OH)	[46]	Abbrev. (R=H)
Adenine	43a	HPMPA	46a	PMEA
2-Aminoadenine	43b	HPMPDAP	46b	PMEDAP
2-Methyladenine	43c		---	
2-Methylthioadenine	43d		46d	
N6-Dimethyladenine	43e		---	
6-Hydrazinopurine	43f		46f	
6-Hydroxylaminopurine	43g		---	
6-Methylthiopurine	---		46h	
Hypoxanthine	43i	HPMPHx	46i	PMEHx
2-Aminopurine	43j	HPMPMAP	46j	PMEMAP
Guanine	43k	HPMPG	46k	PMEG
Isoguanine	43l		46l	
Uracil	43m	HPMPU	46m	PMEU
Thymine	43n	HPMPT	46n	PMET
Cytosine	43o	HPMPC	46o	PMEC
5-Methylcytosine	43p		46p	
5-Fluorouracil	43q		---	

a) Base... pyrimidin-1-yl or purin-9-yl residue.

In addition to HPMPA [**43a**], its 2-aminoadenine [**43b**], guanine [**43k**] and cytosine [**43o**] derivatives emerged as potent antiviral agents against herpes viruses type 1 and 2, and TK^-mutant strain of HSV-1(32) (Table 4) The same compounds are also active against vaccinia virus. Slightly less, but still marked antiviral activity against these viruses was shown by the 6-hydroxylaminopurine [**43g**] and 6-hydrazinopurine [**43f**] analogs of HPMPA. On the other hand, the isomers of HPMPA (adenin-3-yl or 2-aminopurin-9-yl [**43j**]) as well as its 2-substituted [**43c,d,l**] or N^6-substituted derivative [**43e**] were totally devoid of antiviral activity. Similar strict specificity is encountered in the pyrimidine series in which the marked activity of HPMPC [**43o**] sharply contrasts with the inactivity of 5-methylcytosine derivative [**43p**]. Hypoxanthine [**43i**], uracil [**43m**] or thymine [**43n**] derivatives are virtually inactive.

Table 4. Antiviral activity of HPMP- and PME-derivatives (33)

Compound	Abbrev.[a]	MIC_{50} [b] (μg/ml)			
		HSV-1	HSV-2	VV	TK⁻ HSV-1
43a	HPMPA	2	4	0.7	2
43b	HPMPDAP	10	20	2	10
43c		>400	>400	>400	>400
43d		>400	>400	>400	>400
43e		>400	>400	>400	>400
43f		70	70	20	40
43g		10	20	7	7
43i	HPMPHx	>400	>400	>400	>400
43j	HPMPMAP	100	200	150	>400
43k	HPMPG	7	20	2	7
43l		>400	>400	>400	>400
43m	HPMPU	>400	>400	>400	>400
43n	HPMPT	70	70	300	>400
43o	HPMPC	4	10	4	2
43p		>400	>400	>400	>400
43q		300	300	>400	>300
46a	PMEA	7	7	150	7
46b	PMEDAP	2	0.7	20	1
46d		>400	>400	>400	>400
46f		>400	>400	>400	>400
46h		>400	>400	>400	>400
46i	PMEHx	>400	>400	>400	>400
46j	PMEMAP	70	10	>200	150
46k	PMEG[c]	0.2	0.2	1	0.2
46l		7	7	70	10
46m	PMEU	>400	>400	>400	>400
46n	PMET	>400	>400	>400	>400
46o	PMEC	>400	>400	>400	>400
46p		>300	>400	>400	>300

[a] Abbrev., see Table 3; [b] Minimum inhib. concentration required to reduce virus induced cytopathogenicity in primary rabbit kidney cell cultures by 50%. Average values for three herpes simplex virus type 1 (HSV-1) strains (KOS, F, McIntyre), three HSV-2 strains (G, 196, Lyons) and two thymidine kinase deficient (TK⁻) HSV-1 variants (B2006, VMW-1837). VV : vaccinia virus. [c] Cytotoxic for the host cells at >4μg/ml.

The same four HPMP-compounds, i.e. adenine, 2-aminoadenine, guanine and cytosine derivatives, also proved inhibitory to other DNA viruses. The data summarized in Table 5 (33) demonstrate their superiority with respect to other base-modified derivatives listed in Table 3. All the four compounds mentioned are inhibitory to various strains of varicella-zoster virus, cytomegaloviruses and adenoviruses. The comparison of selectivity indexes

is clearly in favor of HPMPA for VZV and AV, whereas HPMPC is superior to the others in CMV inhibition. HPMPG [**43k**] though very active, is the most cytotoxic compound of this series.
A strikingly similar structural dependence of the antiviral activity was encountered also in the PME-series: in this case, the cytosine derivative [**46o**] (PMEC) turned out to be inactive as an antiviral agent. Thus, only the adenine [**46a**], 2-aminoadenine [**46b**] and guanine [**46k**] derivatives emerged as potent antiviral agents, about equally active against HSV-1 and HSV-2 as their HPMP-counterparts (Table 4).

Table 5. Antiviral activity of HPMP- and PME-derivatives

Compound	Abbrev.	MIC_{50} (µg/ml)[a) [SI][b)		
		VZV[c)	CMV[d)	AV[e)
43a	HPMPA	0.02 [<u>1000</u>]	0.15 [133]	0.33 [<u>167</u>]
43b	HPMPDAP	1 [>200]	2 [>100]	2.8 [45]
43i	HPMPHx	>400 [1]	100 [>4]	>125 [1]
43k	HPMPG	0.07 [143]	0.15 [67]	2.6 [2.7]
43m	HPMPU	300 [>1.3]	25 [16]	>125 [1]
43n	HPMPT	70 [>6]	100 [>4]	>125 [1]
43o	HPMPC	0.25 [200]	0.08 [<u>625</u>]	3.4 [>37]
46a	PMEA	10 [10]	25 [4]	>125 [1]
46b	PMEDAP	2 [20]	10 [4]	>125 [<0.5]
46j	PMEMAP	150 [>2]	200 [>2]	>125 [1]
46k	PMEG	0.05 [50]	0.25 [10]	2.7 [0.93]

a) in human embryonic lung cells; b) selectivity index: ratio of minimum cytotoxic concentration to minimum antiviral concentration; c) average value for two TK^+ varicella-zoster virus (VZV) strains (YS, Oka) and two TK^- VZV strains (07-1, YSR); d) average value for two cytomegalovirus (CMV) strains (Davis, AD169); e) average value for three adenovirus (AV) types (2,3 and 4).

In this series, also the isoguanine (2-hydroxyadenine) derivative [**46l**] displayed marked antiviral activity whereas the isomers of PMEA [**46j**] and the N ^{3}-derivative were much less efficient. The PME-derivatives turned out to be ineffective against adenoviruses (Table 5) and vaccinia virus (Table 4) (the apparent activity of PMEG is counteracted by its high cytotoxicity). Also their potency against VZV and CMV is inferior to that of their HPMP-counterparts.

Another item in which the PME-compounds [**46**] clearly differ from the HPMP-series, is their activity against retroviruses. In murine fibroblast (C3H) cells infected with Moloney MSV, PMEA and its 2-aminoadenine derivative [**46b**] significantly reduced the number of MSV-infected foci at very low concentrations,with selectivity indexes of 200 and 67, respectively. The guanine derivative PMEG and 2-aminopurine analog [**46j**] also exhibited marked effects. However, these compounds are cytotoxic to the host cells (36). None of the additional base modified PME-compounds listed in Table 3 showed any anti-MSV activity at concentrations up to 200µM. In contrast to the PME-derivatives, HPMPA and its congeners were considerably less active against Moloney MSV: the parent compound, as well as HPMPG [**43k**] or the 2-aminopurine derivative [**43b**] were at least 10 times less efficient against MSV-transformation than their PME-counterparts (Table 6).

Table 6. Anti-HIV activity of HPMP- and PME-derivatives and their effect on MSV-induced cell transformation(34)

Compound	Abbrev.	ED_{50} (µM)$^{a,b)}$	
		HIV $^{c)}$	MSV $^{d)}$
43a	HPMPA	>125 [<1]	>8 [<5]
43b	HPMPDAP	>125 [<1]	6 [33]
43i	HPMPHx	>125 [<1]	>200 [1]
43k	HPMPG	>25 [<1]	6 [6.7]
46a	PMEA	2.0 [33]	1 [200]
46d	PMEDAP	1.0 [18]	0.6 [67]
46i	PMEHx	>125 [<5]	>200 [1]
46j	PMEMAP	45.0[>28]	1.21[1.3]
46k	PMEG	3.8 [3]	0.12[2.5]
	ddCyd	0.47 [170]	29 [>6.9]

$^{a)}ED_{50}$: the dose required to achieve 50% protection of the cells against HIV infection (as monitored by cell viability determined 5 days after infection)or to reduce the number of MSV-induced foci by 50%. $^{b)}$in parentheses, selectivity index [SI](see Table 4).$^{c)}$in MT-4 cells;$^{d)}$in C3H cells.

PMEA, PMEG, PMEDAP and PMEMAP proved to be potent inhibitors of HIV replication in MT-4 cells (Table 6) (36). Complete protection was achieved at 5μM with PMEA and PMEDAP. The same structure-activity pattern was observed in ATH8 cells infected with HIV:PMEDAP was the most active antiviral of the group tested, with an ED_{50} value of 1.5μM (36).The potencies of selected PME-derivatives are comparable to those reported for other antivirals, e.g. 2',3'-dideoxyribonucleosides. Since PMEA and/or its analogs will be less apt to undergo degradative processes *in vivo* than the nucleoside antivirals, it might seem important to further pursue them as potential anti-HIV agents.

The antiviral potency *in vitro* of both series of compounds was further confirmed *in vivo* with several animal models: HSV-1 keratitis in rabbits (topical treatment) (33,37), HSV-1 infection in mice (topical, and systemic treatment), HSV-1 and HSV-2 infection in guinea pigs (38) and vaccinia virus infection in mice. The experiments which were performed with HPMPA (23,33) clearly demonstrate a marked protective effect against the manifestations of viral infection. PMEG is reported to have ED_{50} of 0.125 mg/kg (i.p.)against systemic HSV-2 infection in mice (compared with 50 mg/kg for acyclovir)(39). Also, PMEA suppressed tumor formation and associated mortality in mice inoculated with MSV by 90-100% at 20-50 mg/kg/day (40).Interestingly, the combination of PMEA with AZT (Retrovir) results in a greater antiretroviral effect than either of the drugs alone.

Thus, it would seem mandatory to further explore the terapeutic potential of this new class of antivirals for the treatment of viral infections in humans.

Biochemical Studies

The studies on the metabolism and mode of action of HPMPA were performed in human embryonic cells (HEL) and Vero cells, both non-infected and infected with HSV-1 (KOS), with the use of (S)-[^{14}C]HPMPA (41,42). It was demonstrated that the compound penetrates into mock-infected and virus-infected cells where it is phosphorylated to its diphosphate (ribonucleoside 5'-triphosphate analog). The analysis of acid-soluble cell pool further indicated some difference between the extent of HPMPA phosphorylation in virus- and mock-infected cells: the sum of labeled HPMPA mono- and diphosphate isolated from acid-soluble cell pool after 6 h treatment with 15 μM HPMPA was 2-2.5 x higher in virus-infected than mock-infected cells irrespective of the cell line used (41). The intracellular concentration depended on the cell line used; it was approximately 5x higher in HEL than Vero cells and amounted to 18 pmole/10^6 HEL cells.

The phosphorylation of HPMPA is catalyzed by cellular nucleotide kinases. This was also confirmed by *in vitro* phosphorylation of HPMPA in the presence of 100000 g supernatant from L-1210 mouse leukemia cells. This phosphorylation requires the presence of ATP; the reaction is stimulated by the ATP-regenerating system(41,43) GTP, UTP and CTP being less efficient phosphate donors in this reaction (41). Also, PMEA was phosphorylated under these conditions;however,the efficacy of this transformation was approximately 5x lower than that of HPMPA (43)

HPMPA inhibits very efficiently viral (HSV-1)DNA synthesis. The effect depends on the host cell line used. The incorporation of [^{32}P]-orthophosphate into viral DNA of HSV-1-infected Vero cells was completely suppressed if HPMPA was added at a concentration of 500 µM whereas the same effect was achieved in HEL cells only at 0.5µM HPMPA. A similar difference was observed in terms of antiviral activity: the HSV-1 replication in HEL cells was inhibited at HPMPA concentration 1000x lower than that required for virus inhibition in Vero cells (41).

Our experiments also demonstrate that under experiental conditions where viral DNA synthesis is fully inhibited, cellular DNA synthesis de novo is inhibited by only 50%. In Vero cells, HPMPA is significantly incorporated into cellular DNA after 24h treatment with the drug at a concentration which causes 50% inhibition of viral DNA synthesis (and does not influence the cellular DNA synthesis). However, no incorporation into viral or cellular DNA was found in HEL cells under equivalent conditions. The incorporation of HPMPA into cellular DNA in Vero cells is extremely low and does not exceed 0.5 pmole/10^6 cells. It is difficult to discern whether it reflects an internucleotidic process or DNA chain termination. The extremely low intracellular concentration of HPMPA that is sufficient to inhibit viral DNA synthesis also explains why HPMPA incorporation based upon radioactivity falls below the level of detection.

The inhibition of cellular DNA synthesis by HPMPA and PMEA was also followed by the incorporation of [^{14} C]-labelled 2'-deoxythymidine in L-1210 mouse leukemia cell culture. The cytostatic effects of both compounds on L-1210 cells were also determined. Whilst the IC_{50} values amounted to 57µM for HPMPA and 15.5µM for PMEA, cellular DNA synthesis was reduced by 50% at concentrations as high as 185µM HPMPA and 250µM PMEA(43). This discrepancy which is particularly striking in the case of PMEA suggests that mechanisms other than inhibition of DNA synthesis de novo might be involved, at least in the cytostatic activity of the drugs.

HSV-1 DNA Polymerase. The transformation of HPMPA into its diphosphate (a triphosphate analog) and the experimentally proven inhibition of viral DNA synthesis prompted us to investigate the interaction of this diphosphate with HSV-1 DNA polymerase in vitro (44). The enzyme used for that purpose was purified from HSV-1 (KOS) infected Hela cells. The results presented in Table 7 refer to the inhibitory activities of HPMPA diphosphate and the structurally related diphosphates of N-(2-phos-

phonylmethoxyethyl) derivatives of purine and pyrimidine bases.
These results demonstrate that HPMPA diphosphate is a comparatively weak inhibitor of viral DNA polymerase (K_m/K_i value:0.63). The order of inhibitory activity for HSV-1 DNA polymerase in the series of the diphosphates derived from the PME-derivatives [**47**] corresponded to their relative antiviral efficiency against HSV-1. The K_m/K_i values of most compounds of this series indicated a higher efficiency of these inhibitors against the viral enzyme than noted for the HPMPA diphosphate. However, HPMPA is a more efficient inhibitor of HSV-1 replication than PMEA. This discrepancy could possibly be explained by a difference in the extent of transformation to the diphosphate which is higher for HPMPA than for PMEA (vide supra).

Table 7. Inhibition of HSV-1 DNA polymerase by acyclic nucleoside triphosphate analogs [**47,48**]

Compound[a]	Competitive Substrate	K_m (μM)	K_i (μM)	K_m/K_i
PMEApp	dATP	0.73	0.105	6.95
PMEGpp	dGTP	1.12	0.090	12.4
PMECpp	dCTP	0.90	1.27	1.41
PMETpp	dTTP	1.25	1.01	1.24
PMEUpp	dTTP	1.25	5.90	0.21
PMEDAPpp	dATP	0.73	0.029	25.1
HPMPApp	dATP	0.73	1.42	0.51
AZT-TP	dTTP	1.25	327.0	0.004
ddTTP	dTTP	1.25	21.1	0.06

[a]) Abbreviations, see Table 3; pp... diphosphate residue

Ribonucleotide Reductase. Another virus-induced enzyme which plays an important role in the virus replication cycle and affects the viral DNA synthesis is the HSV-1 specific (virus-encoded) ribonucleotide reductase (ribonucleoside 5'-diphosphate reductase). The viral enzyme differs from its eukaryotic counterpart: it is not allosterically regulated by nucleoside 5'-triphosphates (dATP, ATP and dTTP) (45), but most probably regulated by competition of the substrates for the common catalytic site. It was therefore regarded as one of the possible targets for acyclic nucleotide analogs of the HPMP- and PME-type (46). Partially purified viral ribonucleotide reductase free of cellular enzyme (according to the immunological criteria) was isolated from HSV-1 (KOS) infected REF or Vero cells. The data summarized in Table 8 demonstrate significant inhibition of this enzyme by mono- and diphosphates derived from HPMP-[**48**] and PME-derivatives [**47**]. The enzyme is not modulated by natural 2'-deoxy- and ribonucleoside 5'-triphosphates (data not shown).

$$B\text{-}CH_2CH_2OCH_2\overset{O}{\underset{OH}{P}}\text{-}O\text{-}\overset{O}{\underset{OH}{P}}\text{-}O\text{-}\overset{O}{\underset{OH}{P}}\text{-}OH$$

47

$$B\text{-}CH_2\underset{CH_2OH}{CH}\text{-}OCH_2\overset{O}{\underset{OH}{P}}\text{-}O\text{-}\overset{O}{\underset{OH}{P}}\text{-}O\text{-}\overset{O}{\underset{OH}{P}}\text{-}OH$$

48

Table 8. Inhibition of HSV-1 encoded ribonucleotide reductase reduction of ribonucleotide 5'-diphosphates[a)]

Compound[b)]	IC_{50} [μM]		
	CDP	ADP	GDP
PMEA	>2000	ND[d)]	ND
PMEAp	96	320[c)]	86[c)]
PMEApp	4.6	26.0[c)]	25.7[c)]
PMEGp	>2000	ND	ND
PMEGpp	1460	ND	ND
PMEDAPp	180	ND	ND
PMEDAPpp	19.2	ND	ND
PMECp	>2000	ND	ND
PMECpp	1600	ND	ND
PMETp	>2000	ND	ND
PMETpp	11	ND	ND
PMEUp	>2000	ND	ND
PMEUpp	>2000	ND	ND
HPMPAp	480[c)]	>2000[c)]	370[c)]
HPMPApp	0.9[c)]	18.0[c)]	0.5[c)]

[a)] [S]=16 μM; [b)] Abbreviations, see Table 3; p... monophosphate, pp...diphosphate residue; [c)] [S] = 22 μM; [d)] ND... not determined.

Reverse Transcriptase. Several PME-derivatives [**46**] are active against retroviruses (MSV,HIV). Therefore we have also investigated the affinity of the diphosphates [**47**] derived from these compounds towards reverse transcriptase - a key enzyme for retrovirus multiplication (47). The assay was performed with detergent-disrupted AMV retrovirions as the source of the reverse transcriptase and template for the oligo(dT)$_{12-18}$ -primed reverse transcription reaction. The activity of the above mentioned diphosphates was compared with AZT triphosphate and ddTTP as reference compounds. It was found that the diphosphates [**47**] inhibit the enzyme; the inhibition depended upon the character of the base and decreased in the order of 2-aminoadenine > adenine > guanine > cytosine > uracil = thymine. The 2-aminoadenine derivative [**47**] was more potent than either AZT 5'-triphosphate and ddTTP while the adenine derivative (PMEA diphosphate) had approximately the same potency as the two reference compounds (Table 9). The last mentioned compound inhibited both the RNA-dependent and DNA-dependent reactions catalyzed by the reverse transcriptase.

Table 9. Inhibition of AMV reverse transcriptase by acyclic analogs of nucleoside 5'-triphosphate [48]

Compound[a)]	IC_{50} (μM)[b)] 3 min	5 min
PMEDAPpp	0.23	0.18
PMEApp	1.35	1.00
PMEGpp	2.50	2.10
PMETpp	3.60	3.25
PMEUpp	3.90	3.10
AZT-TP	1.05	1.13
ddTTP	1.50	1.00

[a)]Abbreviations, see Footnote of Table 8. [b)]Inhibitor concentration causing 50% depression of labelled NTP incorporation into the growing DNA chain, after a 3 or 5 min incubation time.

Conclusion

Two groups of biologically active acyclic nucleotide analogs which resulted from our structure-activity investigation can formally be represented by a single general formula [49]:

$$B\text{-}CH_2CH(R)\text{-}OCH_2P(O)(OH)_2 \quad \textbf{[46]}\ R = H;\quad \textbf{[43]}\ R = (\underline{S})\text{-}CH_2OH$$

[49]

Nevertheless, the data which are so far available on their biological, and in particular antiviral activity, as well as on their in vitro effects in isolated enzyme systems, suggest that both groups may act by different mechanisms. The narrow structural margin which excludes even the smallest variations of the parent structures, including isosters and carba analogs, points to the importance of the structural features of these molecules and of their mutual arrangement. The comparatively high selectivity for heterocyclic bases containing amino group(s) capable of protonation might be an important lead for the understanding of their behavior in biological systems.

Penetration of these compounds into living cells is limited. This fact perhaps explains certain discrepancies between the activities in isolated enzyme and intact cell systems. Further development in the direction of prodrugs, in particular with protracted action or facilitated transport, might lead to new avenues for therapeutic application. Although the animal experiments point to a comparatively lower toxicity of the adenine series (PMEA, HPMPA), than of the guanine series (PMEG, HPMPG), an increased intracellular concentration of the analogs might well be a disadvantage.

Our structure-activity studies were necessarily not quite exhaustive. It is possible that additional side-chain modified acyclic nucleotide analogs might exhibit antiviral or other properties. A better understanding of the biochemical mode of action of the two classes of acyclic nucleotide analogs might help to foster further developments in this new field.

Literature cited

1. Robins,R.K. Pharm.Res. 1984, 11-18.
2. Scheit,K.H. Nucleotide Analogs;Wiley-Interscience: New York,1980.
3. Holý,A.in Phosphorus Chemistry Directed Towards Biology; Stec,W.J.,Ed.;Pergamon Press:Oxford,1980,pp. 53-64.
4. Holý,A., Rosenberg,I. Collect.Czech.Chem.Commun. 1982, 47, 3447-3463.
5. Rosenberg,I., Holý,A. Collect.Czech.Chem.Commun. 1987, 52, 2572-2588.
6. Rosenberg,I., Holý,A. Collect.Czech.Chem.Commun. 1985, 50, 1507-1513.
7. Vesely,J.,Rosenberg,I.,Holý,A. Collect.Czech.Chem. Commun. 1982, 47, 3464-3469.
8. Vesely,J.,Rosenberg,I.,Holý,A. Collect.Czech.Chem. Commun. 1983, 48, 1783-1787.
9. Holý,A.,Nishizawa,M.,Rosenberg,I.,Votruba,I. Collect.Czech.Chem.Commun. 1987, 52, 3042-3057.
10. Horská,K.,Rosenberg,I.,Holý,A.,Šebesta,K. Collect. Czech.Chem.Commun. 1983, 48, 1352-1357.
11. Horská,K.,Cvekl,A.,Šebesta,K.,Rosenberg,I.,Holý,A. Abstr.14th IUB Congress, Fr 355; Prague,1988.
12. Shannon,w.m. in Antiviral Agents and Viral Diseases of Man; Galasso,G.J., Merigan,T.C., Buchanan,R.A., Eds.; Raven Press: New York, 1984; pp.55-121.
13. Robins,R.K., Revankar,G.R. in Antiviral Drug Development.A Multidisciplinary Approach; De Clercq,E., Walker,R.T.,Eds.; Plenum Press: New York, 1988; pp. 11-36.
14. Holý,A. in Approaches to Antiviral Agents; Harnden, M.R., Ed.; Macmillan Press: Basingstoke, 1985; pp. 101-134.
15. De Clercq,E.,Holý,A. J.Med.Chem. 1979, 22, 510-513.
16. De Clercq,E. Nucleosides&Nucleotides 1985, 4, 3-11.
17. Tolman,R.L., Field,A.K., Karkas,J.D., Wagner,A.F., Germershausen,J., Crumpacker,C., Scolnick,E.M. Biochem.Biophys.Res.Commun. 1985, 128, 1329-1335.
18. Hutchinson,D.W., Naylor,M. Nucleic Acids Res. 1985, 13, 8519-8530.
19. Holý,A. Chemica Scripta 1986, 26, 83-89.

20. Votruba,I., Holý,A., De Clercq,E. Acta Virol. 1983, 27, 273-276.
21. Holý,A., Čihák,A. Biochem.Pharmacol. 1981, 30, 2359-2361.
22. Rosenberg,I., Holý,A. Collect.Czech.Chem.Commun. 1883, 48, 778-789.
23. DeClercq,E., Holý,A., Rosenberg,I., Sakuma,T., Balzarini,J., Maudgal,P.C. Nature 1986, 323, 464-467.
24. Holý,A. Collect.Czech.Chem.Commun. 1978, 43, 3103-3117.
25. Holý,A., Rosenberg,I. Collect.Czech.Chem.Commun. 1987, 52, 2775-2791.
26. Rosenberg,I., Holý,A. Collect.Czech.Chem.Commun. 1988, 53, 2753-2777.
27. Webb,R.R., Martin,J.C. Tetrahedron Letters 1987, 28, 4963-4964.
28. Webb,R.R., Wos,J.A., Bronson,J.J., Martin,J.C. Tetrahedron Letters 1988, 29, 5475-5478.
29. Holý,A., Rosenberg,I. Collect.Czech.Chem.Commun. 1987, 52, 2801-2809.
30. Osterhaus,A.D.M.E.,Groen,J.,De Clercq,E. Antiviral Res. 1987, 7, 221-226.
31. Lin,J.C., De Clercq,E., Pagano,J.S. Antimicrob.Ag. Chemother. 1987, 31, 1431-1433.
32. Baba,M., Konno,K., Shigeta,S., De Clercq,E. Eur.J. Clin.Microbiol. 1987, 6, 158-160.
33. De Clercq,E., Sakuma,T., Baba,M., Pauwels,R.,Balzarini,J., Rosenberg,I., Holý,A. Antiviral Res. 1987, 8, 261-272.
34. Gil-Fernández,C., De Clercq,E. Antiviral Res. 1987, 7, 151-160.
35. Baba,M.,Mori,S.,Shigeta,S.,De Clercq,E. Antimicrob. Ag.Chemother. 1987, 31, 337-339.
36. Pauwels,R.,Balzarini,J.,Schols,D.,Baba,M.,Desmyter, J., Rosenberg,I., Holý,A., De Clercq,E. Antimicrob. Ag.Chemother. 1988, 32, 1025-1030.
37. Maudgal,P.C.,De Clercq,E.,Huyghe,P. Invest.Ophtalmol.Vis.Sci. 1987, 28, 243-248.
38. Hitchcock,M.J.M.,Ghazzouli,I., Tsai,Y.H., Bartelli, C.A., Webb,R.R., Martin,J.C. Abstr.2nd Int.Conf.on Antiviral Research, Williamsburg (USA); 1988.
39. Bronson,J.J.,Kim,C.U.,Ghazzouli,I.,Hitchcock,M.J.M. Martin,J.C.Abstr.2nd Int.Conf.on Antiviral Research Williamsburg (USA) 1988.
40. Balzarini,J., Naesens,L., Rosenberg,I., Holý,A., De Clercq,E.Abstr.Int.Symp.on AIDS,San Marino;1988.
41. Votruba,I., Bernaerts,R., Sakuma,T., De Clercq,E., Merta,A.,Rosenberg,I., Holý,A. Mol.Pharmacol. 1987, 32, 524-529.
42. Sakuma,T., De Clercq,E., Bernaerts,R., Votruba,I., Holý,A. in Frontiers in Microbiology, Martinus Nijhoff Publishers; Dordrecht: 1987, pp.300-304.
43. Veselý,J.,Merta,A.,Votruba,I.,Rosenberg,I., Holý,A. Abstr.14th IUB Congress, Tu 431; Prague,1988.

44. Merta,A.,Votruba,I.,Rosenberg,I.,Otmar,M., Holý,A. Abstr.14th IUB Congress, Tu 433; Prague,1988.
45. Averett,D.R., Lubbers,C., Elion,G.B., Spector,T. J.Biol.Chem. 1983, 258, 9831-9838.
46. Černý,J., Vonka,V.,Votruba,I., Rosenberg,I.,Holý,A. Abstr.14thIUB Congress, Tu 621; Prague,1988.
47. Votruba,I., Trávníček,M., Rosenberg,I., Otmar,M., Holý,A.Abstr.14thIUB Congress, Tu 432; Prague,1988.

RECEIVED February 14, 1989

Chapter 5

Synthesis and Antiviral Activity of Phosphonylmethoxyethyl Derivatives of Purine and Pyrimidine Bases

Joanne J. Bronson[1], Choung Un Kim[1], Ismail Ghazzouli[1], Michael J. M. Hitchcock[1], Earl R. Kern[2], and John C. Martin[1]

[1]Bristol-Myers Company, Pharmaceutical Research and Development Division, 5 Research Parkway, Wallingford, CT 06492–7660
[2]University of Alabama, Birmingham, AL 35294

Acyclic nucleotide analogues having a common (2-phosphonylmethoxy)ethyl (PME) side chain attached to a purine or pyrimidine base were prepared and selected for in vivo studies based on in vitro antiviral activity against retroviruses and herpesviruses. The adenine derivative PMEA (**2**) showed good in vitro activity against human immunodeficiency virus (HIV) and Rauscher murine leukemia virus (R-MuLV), and was more potent in vivo than 3'-azido-3'-deoxythymidine (AZT) in the treatment of R-MuLV infection in mice. PMEA also had a significant antiviral effect in vivo against murine cytomegalovirus. The guanine derivative PMEG (**10**) was exceptionally potent in vitro against herpesviruses. In vivo, PMEG was >50-fold more potent than acyclovir against HSV 1 infection in mice. A number of analogues of PMEG were also prepared, but these derivatives were less potent than PMEG or devoid of antiviral activity.

The report by De Clercq and Holy (1) on the broad-spectrum antiviral activity of (S)-9-((3-hydroxy-2-phosphonylmethoxy)propyl)adenine (HPMPA, **1**) has prompted interest in the synthesis and evaluation of related acyclic nucleotide analogues. Replacement of the adenine base in HPMPA with different purine and pyrimidine bases has provided other HPMP-derivatives with potent and selective anti-DNA virus activity (2). A series of phosphonate derivatives similar to HPMPA, but lacking the hydroxymethyl appendage present on the HPMP side chain, have also shown significant antiviral activity. The adenine derivative in this series, 9-((2-phosphonylmethoxy)ethyl)-adenine (PMEA, **2**) (3), was first reported to have in vitro activity against a murine retrovirus (1). Further studies have shown that PMEA and related PME-derivatives are effective inhibitors of human immunodeficiency virus (HIV), the human retrovirus which causes acquired immunodeficiency syndrome (AIDS) (4). In addition, members of the PME-series have shown in vitro potency similar to the HPMP-derivatives against DNA viruses such as herpes simplex viruses (2).

0097–6156/89/0401–0072$06.00/0

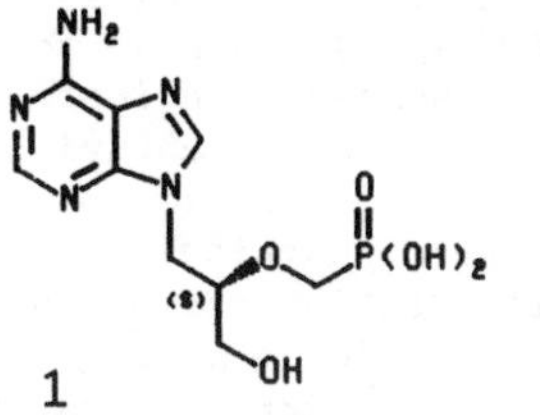

1 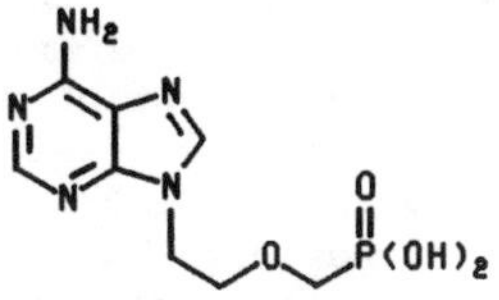 2

A further aspect of the antiviral activity of HPMPA and related phosphonate derivatives is their inhibitory effect on DNA viruses that lack viral thymidine kinase (TK) activity, including TK-deficient strains of herpes simplex virus (HSV), as well as viruses that do not encode TK such as cytomegalovirus (CMV). Activity against CMV is particularly important since this virus is one of the leading causes of opportunistic infection in AIDS patients, and there is currently no therapy approved in the U.S. for treatment of CMV infection. These types of viruses are generally resistant to nucleoside analogues such as acyclovir (ACV) that are dependent on initial phosphorylation by viral TK in order to exert an antiviral effect. For example, the mechanism of action of ACV against HSV 1 involves preferential monophosphorylation by HSV-encoded TK and then further phosphorylation by cellular kinases to a triphosphate analogue. It is the triphosphate metabolite of ACV which acts as an inhibitor of viral DNA polymerase and consequently viral replication (5). By contrast, the potent antiviral activity of HPMP- and PME- derivatives against CMV and TK^- strains of HSV (2) demonstrates that their mode of action is not dependent on virus-specified TK. These phosphonate derivatives act as stable monophosphate equivalents and can therefore bypass the virus-dependent phosphorylation step. In biochemical studies on the mechanism of action of these nucleotide analogues, HPMPA was shown to be converted directly by cellular kinases to di- and triphosphate analogues (6). In addition, HPMPA was shown to inhibit viral DNA synthesis at a concentration much lower than required for inhibition of cellular DNA synthesis (6).

While substitution of a phosphonic acid group for a phosphate monoester is a common strategy in designing metabolically stable phosphate equivalents (7), HPMPA and related compounds are unique nucleotide analogues in that they are phosphonylmethyl ether derivatives [$O-CH_2-P(O)(OH)_2$] rather than simple alkyl phosphonates [$R-CH_2-P(O)(OH)_2$]. The electron-withdrawing oxygen substituent adjacent to the phosphonate group may serve an important role by providing a site for binding to target enzymes and influencing the electronic nature of the phosphorous moiety. For example, the effect of the electron-withdrawing substituent on the degree of dissociation can be seen by comparison of the second pK_a values for different phosphorous derivatives (7,8). For the phosphonylmethyl ether derivative PMEA, pK_{a2} is 6.8, closely resembling that of phosphate esters [$R-O-P(O)(OH)_2$] which are in the range of 6.5 - 7.0; alkyl phosphonates are less acidic with pK_{a2} values of 7.7 - 8.2. The isoelectronic nature of the phosphorous moiety may be an important factor in the potent and broad-spectrum antiviral activity demonstrated by members of this class of nucleotide analogues.

Following our initial interest in HPMPA (9), we have continued the synthesis and evaluation of related phosphonylmethyl ether derivatives in order to determine their potential as antiviral agents. Our work on the PME-series of nucleotide analogues is the focus of this chapter. These compounds are of particular interest because of their antiviral activity against retroviruses and certain DNA viruses. Synthetic routes capable of providing multigram quantities of various PME purine and pyrimidine derivatives are described along with the preparation of analogues of the guanine derivative PMEG, which was found to be the most potent antiviral agent in the PME-series. In vitro antiviral activity and extended in vivo evaluation of selected phosphonylmethyl ether derivatives in murine infection models are presented.

Synthesis of Phosphonylmethyl Ether Derivatives

Phosphonylmethoxyethyl (PME) Purine and Pyrimidine Derivatives. For the synthesis of PME-derivatives, a general approach was employed which involves coupling of the appropriate heterocyclic base with a side-chain fragment bearing the phosphonylmethyl ether group. Synthesis of the key intermediate **5** was achieved in a four-step process starting with ring opening of 1,3-dioxolane by treatment with acetyl chloride in the presence of zinc chloride (10). Arbuzov reaction of the resulting chloromethyl ether **3** with triethyl phosphite at 100 °C provided phosphonylmethyl ether **4a** in 60% overall yield after distillation. Removal of the acetate group was effected under acidic conditions (conc. hydrochloric acid, aqueous ethanol), and the resulting alcohol was treated with mesyl chloride and triethylamine to furnish the mesylate **5a** in over 90% yield from **4**; both the alcohol and mesylate are used without purification. Diisopropyl phosphonate ester **5b** was prepared in a similar manner using triisopropyl phosphite in the Arbuzov reaction. Reaction of **3** with the sodium salt of dimethyl phosphite provided access to the corresponding methyl ester derivative **5c**.

OAc, O, Cl — OAc, O, $P(OR)_2$ — OMs, O, $P(OR)_2$

3 4 5

a, R = Et; b, R = *i*-Pr; c, R = Me

Synthesis of the adenine derivative PMEA (**2**) was achieved by coupling of the side-chain derivative **5** with adenine followed by deprotection of the phosphonic acid moiety. The alkylation reaction was carried out by treatment of the sodium salt of adenine with mesylate **5a** in DMF at 100 °C to provide the diethyl ester of PMEA (**6a**) in 55% yield. Formation of 9-ethyladenine as a byproduct was also observed; this product arises from competitive reaction of the adenine salt with the phosphonate ethyl ester and was separated from

6a by careful chromatography. On larger scales, it was more convenient to use the iso-propyl phosphonate derivative **5b** in the coupling reaction, since formation of an alkylated adenine byproduct is minimized and purification is simpler in the absence of the comigrating byproduct; a similar yield of the coupled product **6b** was obtained. Reaction of **6a** or **6b** with bromotrimethylsilane (11) in acetonitrile, followed by concentration and then aqueous hydrolysis of the resulting silylated intermediate, afforded PMEA as a zwitterionic crystalline solid in 90-95% yield.

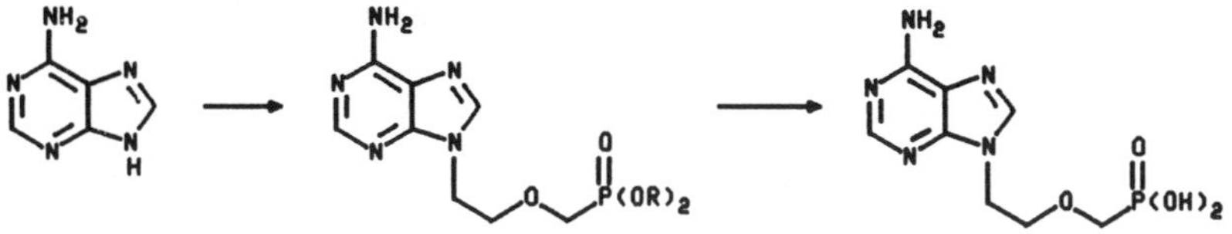

a, R = Et; b, R = i-Pr

Formation of regioisomeric N-9 and N-7 substitution products is typically observed in alkylation reactions of guanine and related derivatives. In initial efforts directed toward the synthesis of the guanine derivative PMEG (**10**), reaction of mesylate **5a** with N^2-acetyl guanine was found to give a 2:1 mixture of alkylated products with the N-7 isomer predominating. The desired N-9 alkylated product was isolated in only 15% yield. Improved results were obtained when 6-O-(2-methoxyethyl)guanine or 6-O-benzylguanine was employed in the coupling reaction: the ratio of N-9/N-7 isomers was 1-2:1, and the N-9 isomer could be isolated in 30-45% yield. The high degree of regioselectivity (up to 15:1 ratios of N-9/N-7 isomers) reported for alkylation of these 6-O-protected guanine derivatives with 4-bromobutyl acetate (12) was not observed for coupling with the mesylate **5a**. Further investigation showed 2-amino-6-chloropurine to be more useful in our alkylation procedure (13). Reaction of this guanine synthon with sodium hydride and then **5a** gave a >6:1 ratio of the alkylated products **7** and **8**, with the desired N-9 isomer **7** isolated as the major product in 55-65% yield.

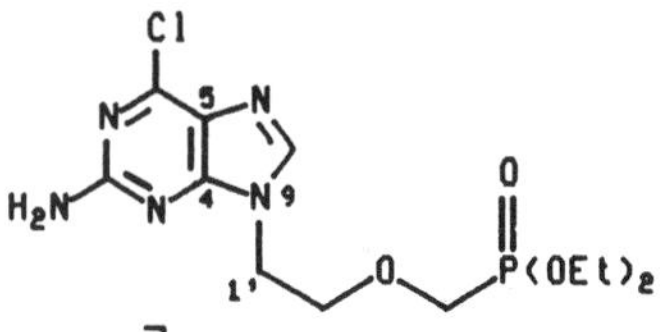

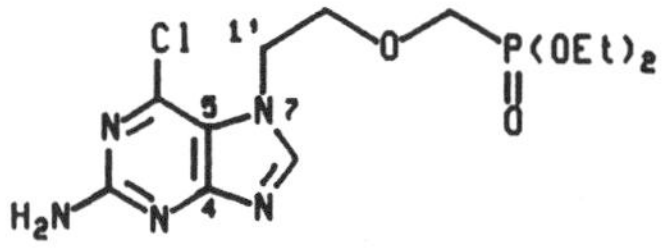

The structure assignments for **7** and **8** were based on ^{1}H and ^{13}C NMR spectroscopic data (14) and confirmed by a 2D NMR experiment: for the N-9 isomer **7**, a three-bond coupling interaction was observed between C-4 of the purine base and the protons at C-1' on the side chain, while this long range coupling was seen between C-5 and H-1'

for the N-7 alkylated product **8**. Conversion of **7** to PMEG (**10**) was achieved in 70-80% overall yield by deesterification with bromotrimethylsilane in DMF to provide the phosphonic acid **9**, followed by hydrolysis of the chloro group with aqueous acid at reflux.

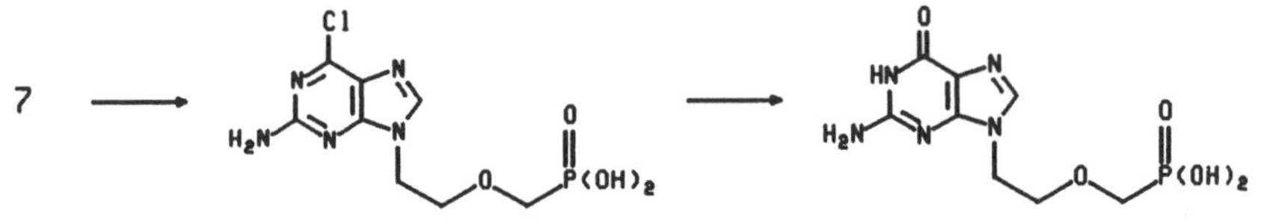

7 → 9 → 10

The 2-amino-6-chloropurine intermediate **7** proved useful for the synthesis of other PME-purine derivatives. For example, removal of the chloro group via transfer hydrogenation (20% $Pd(OH)_2$ on carbon, cyclohexene/ethanol) and then standard deprotection of the phosphonate ester provided the 2-aminopurine derivative **11** (PMEMAP) in 80% overall yield. The 2,6-diaminopurine derivative **12** (PMEDAP) was prepared via displacement of the chloro group with sodium azide, followed by reduction of the azido group and deesterification of the phosphonate ester. In this sequence, use of the diisopropyl ester derivative of **7** was found to give better yields in the reaction with sodium azide, probably because of decreased side reactions at the phosphonate ester moiety. The overall yield for PMEDAP (**12**) was 50% from the 2-amino-6-chloropurine intermediate **7**.

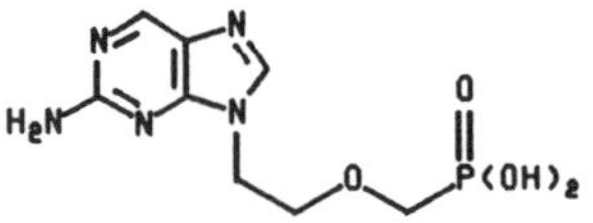

11 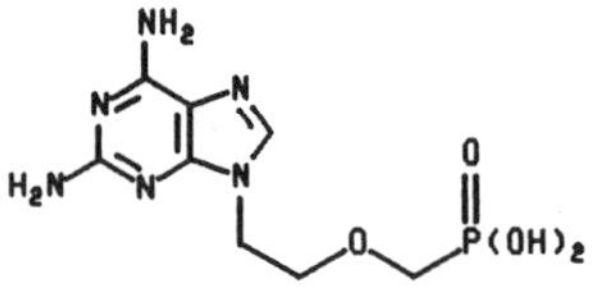 12

For synthesis of the pyrimidine derivative PMEC (**15**), coupling of cytosine with the side-chain mesylate **5** was required. This transformation was carried out by treatment of a mixture of **5a** and cytosine with potassium carbonate in DMF to afford a 4:1 ratio of N- and O-alkylated products **13** and **14**. The N-1 isomer **13** was isolated in 45% yield after separation from the less polar isomeric product by chromatography. Conversion of **13** to PMEC was achieved in 80% yield by deprotection using bromotrimethylsilane.

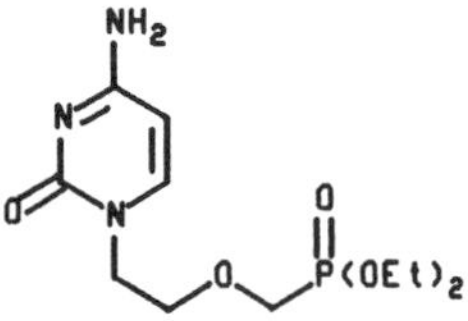

13 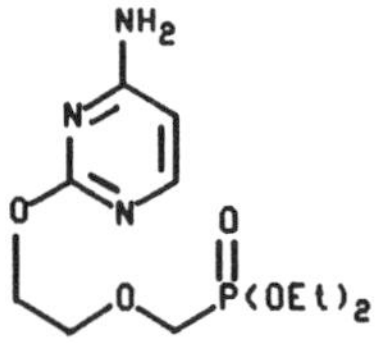 14 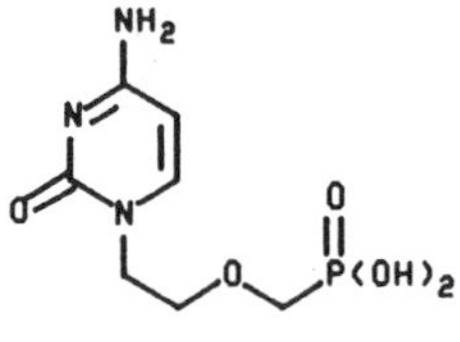 15

Structure assignments for **13** and **14** were based on comparison of their NMR spectra with those for known cytosine derivatives. For example, the chemical shifts of the carbons on the pyrimidine ring are nearly identical for **13** and cytidine (Table I), while for the O-alkylated isomer **14**, the pattern of the aromatic carbon peaks is substantially different. For the product assigned as the N-isomer **13**, the chemical shift of C-1' is also upfield relative to that for the O-isomer **14** (48 vs. 66 ppm). Furthermore, a 2D NMR experiment showed three-bond coupling between C-1' and the proton at C-6 of the cytosine base for the N-alkylated isomer **13** but not for the isomeric product **14**.

Table I. ^{13}C NMR Data for Cytosine Derivatives

Compound	Chemical Shift (δ)			
	C-2	C-4	C-5	C-6
13 (N-1 isomer)	156	166	93	146
14 (O-2 isomer)	165	165	100	157
cytidine	157	166	95	143

Analogues of 9-(2-Phosphonylmethoxy)ethylguanine (PMEG). The finding that the guanine derivative PMEG (**10**) is an exceptionally potent antiherpesvirus agent (vide infra) has prompted efforts to synthesize a number of guanine derivatives related to this lead compound. The synthesis of monoesters of PMEG was of interest to determine the effect of changing the ionic character of the phosphorous moiety. These derivatives might exert a direct antiviral effect or be metabolized to PMEG. Preparation of the monoethyl ester of PMEG (**16**) was accomplished in 75% yield by treatment of 2-amino-6-chloropurine derivative **7** with sodium hydroxide solution at reflux for 2 hours to effect cleavage of one of the ester groups and hydrolysis of chloro group on the purine base. Under these conditions, formation of the diacid occurred only to a small extent (<0.2% by HPLC), although up to 1% PMEG was obtained after prolonged heating (>8 hours) of the reaction mixture. The monoisopropyl and monomethyl esters of PMEG (**17** and **18**) were prepared from the corresponding diester derivatives analogous to **7**. Alternatively, diesters of PMEG served as precursors to the desired monoesters. The required diesters were prepared by coupling of the sodium salt

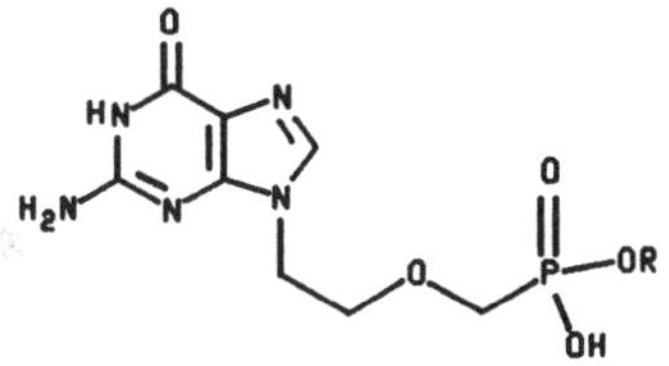

16: R = Et

17: R = i-Pr

18: R = Me

of 6-O-benzylguanine with mesylate **5**, followed by removal of the benzyl group under transfer hydrogenation conditions. The desired PMEG monoester derivatives were then obtained upon treatment of the diester with 1 N sodium hydroxide at room temperature for 1 h.

Analogues of PMEG with longer acyclic side chains (15) are of interest because of their similarity to phosphorylated metabolites of acyclovir. For example, 9-(3-phosphonylmethoxy)propylguanine (PMPG, **22**) can be considered an isosteric analogue of ACV monophosphate since the same number of atoms separate the guanine base and the phosphorous group. The synthesis of PMPG is representative of the procedure used to prepare the longer chain (butyl, pentyl, hexyl, and heptyl) derivatives. The side chain required for PMPG was prepared by chloromethylation [paraformaldehyde, hydrogen chloride (g)] of 3-bromopropanol (**19**), followed by reaction with the sodium salt of diethyl phosphite to afford bromide **20** in 75% yield. Coupling of 6-O-benzylguanine with **20** gave a 2:1 mixture of N-9/N-7 isomeric products; reaction of **21** with TMSBr in DMF effected cleavage of the phosphonate and guanine protecting groups to provide PMPG (**22**) in 30% overall yield from bromide **20**.

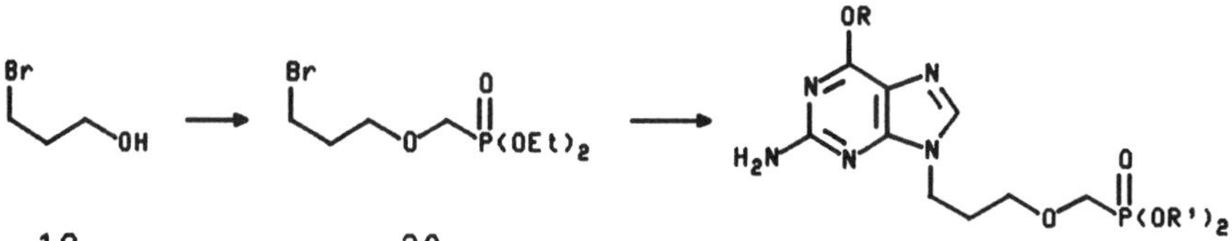

Br OH → Br O P(OEt)$_2$ → OR N N H_2N N N O O P(OR')$_2$

19 20

21: R = CH_2Ph, R' = Et

22: R = R' = H

To determine the effect of increased steric bulk adjacent to the phosphorous group, α-methyl and α,α-dimethyl derivatives of PMEG (**23** and **24**) were prepared. The synthetic approach to these compounds is based on the ability of the phosphonate group to stabilize an α-carbanion which can then undergo substitution reactions with electrophiles such as methyl iodide. The key

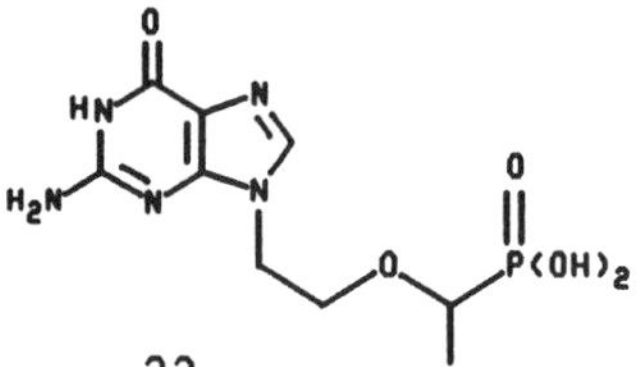

O HN N H_2N N N O O P(OH)$_2$ 23

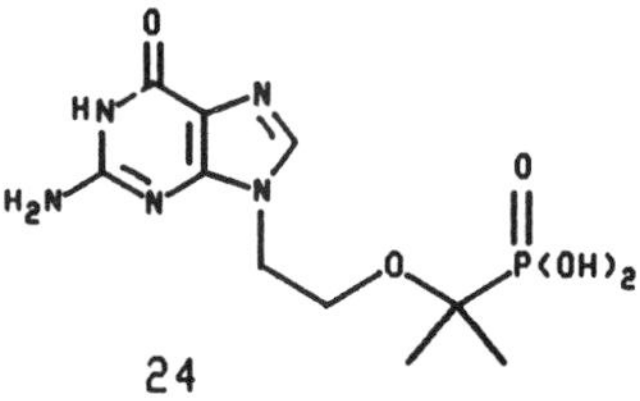

O HN N H_2N N N O O P(OH)$_2$ 24

substrate for the alkylation reaction, the silyl ether **25b**, was prepared by treatment of alcohol **25a** with tert-butyldimethylsilyl chloride and triethyl amine. Note that **25a** is an intermediate in the PME side-chain synthesis. Reaction of **25b** with 1.2 equivalents of sec-BuLi (16) and then methyl iodide at -78 °C gave the alkylated product **26b** in 85-90% yield. Repetition of this process provided the dimethylated phosphonate **27b** in 90-95% yield. Removal of the silyl protecting group in **26b** (acetic acid, aq THF), followed by mesylation of the resulting alcohol, provided the required side-chain derivative **28** in 80-85% yield; the α,α-dimethyl mesylate **29** was obtained in a similar yield by the same sequence. The syntheses of α-methyl PMEG (**23**) and α,α-dimethyl PMEG (**24**) were then completed using the chemistry previously described for the synthesis of PMEG.

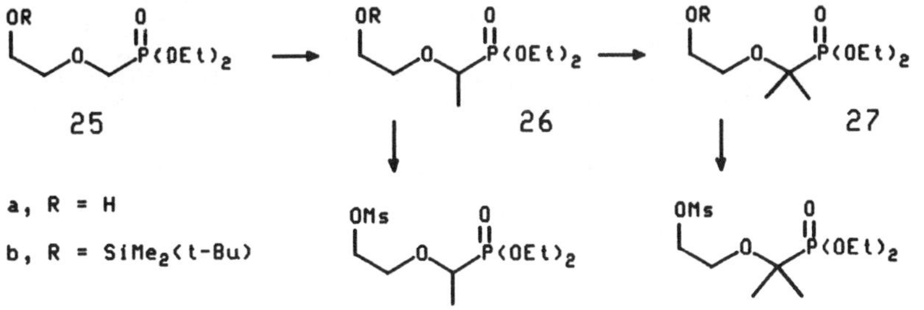

25 → 26 → 27

26 → 28; 27 → 29

a, R = H

b, R = $SiMe_2(t\text{-}Bu)$

Antiviral Activity of Phosphonylmethyl Ether Derivatives

In Vitro and In Vivo Antiviral Activity of PMEA. The original report by De Clercq and Holy on the broad-spectrum anti-DNA virus properties of HPMPA included a description of the inhibitory effect of PMEA on a murine retrovirus, Moloney sarcoma virus (MSV) (1). In a subsequent report, PMEA was shown to have in vitro antiviral activity against DNA viruses as well, although it was less potent than HPMPA (2). More recently, PMEA has been shown to have significant antiviral activity in vitro against HIV, and to be the least toxic of the PME-derivatives having anti-HIV activity (4). The anti-retroviral activity of PMEA has been demonstrated further in vivo using a retrovirus infection model in mice: PMEA was found to be more effective than AZT in suppression of tumor formation caused by infection with Moloney sarcoma virus (17).

Our interest in PMEA has focused on further evaluation to determine its utility as an antiviral agent, using in vitro results as a guide for selection of in vivo experiments. The antiviral effect of PMEA was examined in vitro (Table II) against human immunodeficiency virus (HIV), Rauscher murine leukemia virus (R-MuLV), herpes simplex virus (HSV) types 1 and 2, and human cytomegalovirus (HCMV). PMEA showed good anti-retroviral activity as demonstrated by its inhibitory effect on HIV and R-MuLV, although it was less potent than AZT against both viruses. PMEA was less potent against herpes simplex virus than acyclovir (ACV) or ganciclovir (DHPG), but showed similar activity to DHPG against HCMV.

The Rauscher murine leukemia virus model was chosen to study the antiviral activity of PMEA compared with AZT against retrovirus infection in vivo (Table III). The antiviral effect of each drug was shown by significant inhibition of the splenomegaly caused by

Table II. In Vitro Antiviral Activity of PMEA Against Retroviruses and Herpesviruses

	ID_{50} (μg/mL)				
Compound	HIV	R-MuLV	HSV 1	HSV 2	HCMV
PMEA (2)	3.0	0.05	21.0	9.4	2.7
ACV	-	-	0.5	0.3	38
DHPG	-	-	0.23	0.94	1.8
AZT	0.1-0.5	0.002	-	-	-

- The 50% inhibitory dose was determined in HIV-infected CEM cells by reduction in p24 gag protein (enzyme capture assay). The XC assay (18) was used to determine 50% inhibition of syncitia formation in cells infected with Rauscher MuLV. In HSV-infected vero cells and HCMV-infected MRC-5 cells, the ID_{50} was determined by plaque-reduction assays. Virus strains: HSV 1, BW^s; HSV 2, G; HCMV, AD-169.

virus infection as compared with placebo-treated infected control. At a dose of 100 mg/kg per day, administration of PMEA resulted in almost complete inhibition of the development of splenomegaly. A similar effect was seen with AZT at this dose, although PMEA was more effective than AZT in preventing induction of the disease at the lower dose of 40 mg/kg per day. These results are in contrast with the in vitro data which showed PMEA to be less effective than AZT against this murine retrovirus.

Table III. In Vivo Antiviral Efficacy of PMEA and AZT against Rauscher MuLV (I.P. Administration)

Compound	Dose (mg/kg/day)	Ave. Spleen Wt. (g)	% Inhibition of Splenomegaly
PMEA	100	0.14 ± 0.03	99*
	80	0.18 ± 0.11	96*
	40	0.36 ± 0.14	85*
AZT	100	0.22 ± 0.12	94*
	80	0.58 ± 0.53	71*
	40	0.64 ± 0.47	67*
Placebo Control		1.65 ± 0.80	
Uninfected Control		0.12 ± 0.02	

- Mice were infected i.p. with Rauscher murine leukemia virus. Treatment was initiated 4 h post-infection and continued BID for 30 consecutive days. * = p value < 0.05.

Evaluation of the in vivo antiviral activity of PMEA against CMV was also of interest since CMV is a common opportunistic viral infection in AIDS patients. A murine CMV (MCMV) infection model was used in these studies (Table IV) (19), since there is no animal model for human CMV infection. The antiviral effect of the compound was shown by a significant increase in survival time as compared with untreated and placebo-treated controls. For comparison, data is shown from a similar experiment using DHPG. PMEA provided almost complete protection upon systemic administration at a dose of 100 mg/kg per day when given 6 h post-infection: this regimen resulted in 93% survival of the infected animals. Even when treatment with PMEA at this dose was delayed up to 48 h post-infection, a significant antiviral effect was observed. DHPG provided similar protection at a dose of 33.3 mg/kg per day. The good in vitro antiviral activity of PMEA against both HIV and human CMV, coupled with its in vivo potency in murine models of these infections, make it an especially exciting candidate as an antiviral agent.

Table IV. In Vivo Antiviral Efficacy of PMEA or DHPG against Murine Cytomegalovirus Infection (I.P. Administration)

Compound	Dose (mg/kg/day)	Treatment Delay:	% Survival 6 h	24 h	48 h
PMEA	100		93*	60*	67*
	20		67*	47*	27
	4		20	13	0
Placebo Control			-	7	-
Untreated Control			33		
DHPG	33.3		93*	67*	33
	11.2		67*	80*	73*
	3.8		53	13	0
Placebo Control			-	20	-
Untreated Control			27		

- Mice were inoculated i.p. with MCMV. Treatment initiated at the indicated times and was given BID for 5 consecutive days. * = p value < 0.05.

Antiviral Activity of PME-Derivatives Against Herpesviruses. The PME-purine and pyrimidine derivatives were evaluated for their in vitro antiviral activity against HSV types 1 and 2, a thymidine kinase deficient strain of HSV 1, and human CMV. The results summarized in Table V show the in vitro potency of each compound expressed as the concentration required for 50% inhibition of viral replication. The guanine derivative PMEG exhibited significant antiviral activity against each virus and was the most potent of the phosphonate derivatives tested. PMEG exerted a slightly greater antiviral effect than ACV or DHPG against HSV 1 and 2, but was substantially more potent than these nucleoside analogues against a thymidine kinase deficient strain of HSV 1, indicating that the antiviral activity of PMEG is not dependent on viral TK. Against HCMV, PMEG was 20-fold more potent than ganciclovir. The purine derivative PMEDAP also showed good antiviral activity, although it was somewhat less potent than the control compounds. PMEMAP and PMEC exhibited no antiviral effect against HSV type 1 and 2. The relative in vitro antiviral activities of the PME-derivatives are in agreement with the results reported by De Clercq and Holy (2).

Table V. Comparison of the In Vitro Antiviral Activity of PME-Derivatives

	ID_{50} (µg/mL)			
Compound	HSV 1	HSV 1 (TK^-)	HSV 2	HCMV
PMEG (**10**)	0.08	0.04	0.06	0.09
PMEMAP (**11**)	>100	-	>100	-
PMEDAP (**12**)	2.4	-	3.2	5.0
PMEC (**15**)	>100	-	>100	-
ACV	0.5	4.3	0.3	38
DHPG	0.23	>10	0.94	1.8

- The 50% inhibitory dose was determined by plaque reduction assays in HSV-infected vero cells and HCMV-infected MRC-5 cells. Virus strains: HSV 1, BW^S; HSV 1 (TK^-), Z826; HSV 2, G; HCMV, AD-169.

Based on its exceptional in vitro potency, PMEG was selected for further evaluation in vivo against herpes simplex virus infection in mice and compared to treatment with acyclovir. The antiviral effect of each compound was indicated by a significant reduction in mortality and increase in mean survival time as compared with placebo control. Against HSV 1 systemic infection in mice (Table VI), PMEG exhibited a significant antiviral effect upon intraperitoneal (i.p.) administration at a dose of only 0.125 mg/kg per day, while a dose of 10 mg/kg per day of ACV was required for a similar response. At higher doses, PMEG afforded complete protection, although toxicity was observed at doses above 5 mg/kg per day. The therapeutic utility of PMEG is clearly demonstrated by its efficacy over a range of doses and its substantial antiviral effect at a dose >40-fold lower than where toxicity was observed.

PMEG had a similar antiviral effect in the treatment of HSV 2 systemic infection in mice (Table VII). As in the HSV 1 infection model, PMEG was much more potent than acyclovir and exhibited antiviral activity at a dose 10-50-fold lower than where toxicity was observed. Administration of PMEG even at a dose of 0.125 mg/kg per day resulted in a substantial reduction in mortality. The greater in vivo potency of PMEG compared with acyclovir is in contrast with the in vitro data which show PMEG to be only slightly more potent against both HSV 1 and 2.

Table VI. In Vivo Antiviral Efficacy of PMEG against HSV 1 Systemic Infection in Mice (I.P. Administration)

Dose	PMEG		Acyclovir	
(mg/kg/day)	% survival	MST	% survival	MST
10	9	10.8	50*	14.1*
5	100*	21.0*	17	11.3*
1.25	100*	21.0*	0	7.8
0.50	100*	21.0*		
0.25	80*	19.1*		
0.125	60*	17.2*		
0.031	10	8.9		
Placebo Control	0	7.3		

- Mice were inoculated i.p. with HSV 1 (HL-34 strain). Treatment was initiated 3 h post-infection and continued BID for five consecutive days. MST = mean survival time; experiment was terminated at day 21. * = p value < 0.05.

Table VII. In Vivo Antiviral Efficacy of PMEG against HSV 2 Systemic Infection in Mice (I.P. Administration)

Dose	PMEG		Acyclovir	
(mg/kg/day)	% survival	MST	% survival	MST
50			40	13.0*
25			20	13.0*
10	0	7.6		
5	92*	20.1*		
1	100*	21.0*		
0.25	90*	20.3*		
0.125	60*	17.2*		
0.031	20	12.1		
Placebo Control	10	9.0		

- Mice were inoculated i.p. with HSV 2 (G strain). Treatment was initiated 3 h post-infection BID for 5 days. MST = mean survival time; experiment was terminated at day 21. * = p value < 0.05.

Antiviral Activity of PMEG Analogues Against Herpesviruses. PMEG monoesters, longer chain phosphonylmethoxy(alkyl) PMEG analogues, and α-substituted PMEG derivatives were evaluated for their in vitro antiviral effect against herpes simplex viruses (Table VIII). The ester derivatives were less potent than PMEG, although monomethyl PMEG (**18**) showed antiviral activity comparable to acyclovir and ganciclovir. Overall, there was a trend of decreasing antiviral activity with increasing size of the ester group: the monoethyl ester of PMEG (**16**) was less potent than **18**, and *iso*-propyl PMEG (**17**) showed only weak antiviral activity. The results also showed that modifications in the acyclic PME-backbone resulted in a decrease in antiviral activity compared with PMEG. PMPG (**22**), which has an additional carbon in the alkyl side chain and is an isosteric analogue of acyclovir monophosphate, showed no activity against herpes simplex viruses; similar results were obtained for the longer chain derivatives. Substitution α to the phosphonate group was also not tolerated: α-methyl PMEG (**23**) showed only weak activity against HSV 2 and the α,α-dimethyl derivative **24** was devoid of antiviral activity.

Table VIII. Comparison of the In Vitro Antiviral Activity of Analogues of PMEG

Compound	ID_{50} (μg/mL)	
	HSV 1	HSV 2
A. Ester Derivatives of PMEG		
Monoethyl PMEG (**16**)	5	5
Monomethyl PMEG (**18**)	1.6	1.2
Monoisopropyl PMEG (**17**)	75	35
B. Longer Chain Derivatives		
PM(Propyl)G (**22**)	>100	>100
PM(Butyl, Pentyl, Hexyl, or Heptyl)G	>100	>100
C. α-Substituted PMEG Derivatives		
α-Me PMEG (**23**)	>100	48
α-diMe PMEG (**24**)	>100	>100
PMEG	0.08	0.06
ACV	0.5	0.3
DHPG	0.23	0.94

- The 50% inhibitory dose was determined by plaque reduction assays in HSV-infected vero cells. Virus strains: HSV 1, BW^S; HSV 2, G.

The ester derivatives **16** and **18** were considered for further evaluation based on their in vitro activity against HSV 1 and 2. In preliminary studies comparing the toxicity of the ester derivatives, it was found that monomethyl PMEG (**18**) had similar toxicity to PMEG, while the ethyl ester **16** was less toxic. To determine whether an in vivo therapeutic advantage could be obtained with an ester derivative of PMEG, ethyl ester **16** was selected for evaluation in the treatment of HSV 2 systemic infection in mice (Table IX). The results showed that while monoethyl PMEG provided substantial protection at doses of 50 and 100 mg/kg per day, at higher doses, a toxic effect was seen before complete protection was achieved. As in other in vivo studies, the efficacy of this phosphonate derivative was greater than indicated from its in vitro antiviral activity relative to acyclovir. Whether **16** acts as a prodrug of PMEG or exerts an antiviral effect without cleavage of the ester group is a subject of ongoing studies.

Table IX. In Vivo Antiviral Efficacy of Monoethyl PMEG against HSV 2 Systemic Infection in Mice (I.P. Administration)

Dose (mg/kg/day)	Monoethyl PMEG		Acyclovir	
	% survival	MST	% survival	MST
200	84*	19.9*	80*	20.1*
100	92*	20.6*	67*	19.2*
50	80*	19.8*	38*	17.1*
25	46*	15.8*	17	13.8*
12.5	25	13.4*		
6.25	13	11.8*		
Placebo Control	10	10.2		

- Mice were inoculated i.p. with HSV 2 (G strain). Treatment was initiated 3 h post-infection and continued BID for 5 consecutive days. MST = mean survival time; experiment was terminated at day 21. * = p value < 0.05.

Summary

The phosphonate derivatives PMEA and PMEG have been identified from our studies as promising candidates for further evaluation in the treatment of retrovirus and herpesvirus infections. PMEA was of particular interest because of its good in vitro activity against HIV, the human retrovirus responsible for AIDS, and CMV, a common cause of opportunistic infection in AIDS patients. The antiviral effect of PMEA against these viruses was demonstrated in vivo using murine infection models. Although in vitro results indicated that PMEA was less potent against retroviruses than AZT, administration of PMEA at a dose of 40 mg/kg per day to mice infected with Rauscher murine leukemia virus provided greater protection than AZT given at the same dose. PMEA also showed good in vivo activity against murine CMV, giving significant protection at a dose of 100 mg/kg per day.

The guanine derivative PMEG was the most potent of all the phosphonates evaluated for in vitro antiviral activity against herpesviruses. The high potency of PMEG was demonstrated in vivo against HSV type 1 and 2 infection in mice: in both cases, PMEG was significantly more potent than ACV and showed substantial activity at a dose >40-fold lower than where toxicity was seen. A number of guanine derivatives related to PMEG were prepared; however, modifications of the PMEG skeleton generally led to a decrease or complete loss of antiviral activity. Monomethyl and monoethyl ester derivatives of PMEG showed some promise, having good in vitro activity against herpes simplex virus types 1 and 2. In vivo, only the monoethyl ester was less toxic than PMEG; however, it was much less potent against HSV 2 infection in mice and provided no improvement over the therapeutic index of PMEG. Both PMEA and PMEG will be the subject of further investigations into the potential of the phosphonate derivatives as antiviral agents, with PMEA being a particularly promising candidate for antiviral therapy in AIDS patients.

Literature Cited

1. De Clercq, E.; Holy, A.; Rosenberg, I.; Sakuma, T.; Balzarini, J.; Maudgal, P. C. Nature 1986, 323, 464-467.
2. De Clercq, E.; Sakuma, T.; Baba, M.; Pauwels, R.; Balzarini, J.; Rosenberg, I.; Holy, A. Antiviral Res. 1987, 8, 261-272.
3. Holy, A.; Rosenberg, I. Collect. Czech. Chem. Commun. 1987, 52, 2801-2809.
4. Pauwels, R.; Balzarini, J.; Schols, D.; Baba, M.; Desmyter, J.; Rosenberg, I.; Holy, A.; De Clercq, E. Antimicrob. Agents Chemother. 1988, 32, 1025-1030.
5. Elion, G. B.; Furman, P. A.; Fyfe, J. A.; de Miranda, P.; Beauchamp, L.; Schaeffer, H. J. Proc. Natl. Acad. Sci. USA 1977, 74, 5716-5720.
6. Votruba, I.; Bernaerts, R.; Sakuma, T.; De Clercq, E.; Merta, A.; Rosenberg, I.; Holy, A. Mol. Pharmacol. 1987, 32, 524-529.
7. Engel, R. Chem. Rev. 1977, 77, 349-367.
8. Blackburn, G. M.; Eckstein, F.; Kent, D. E.; Perree, T. D. Nucleosides Nucleotides 1985, 4, 165-167.
9. Webb, R. R.; Martin, J. C. Tetrahedron Lett. 1987, 28, 4963-4964.
10. Bailey, W. F.; Rivera, A. D. J. Org. Chem. 1984, 49, 4958-4964.
11. McKenna, C. E.; Schmidhauser, J. J. Chem. Soc., Chem. Commun. 1979, 739.
12. Kjellberg, J.; Liljenberg, M.; Johansson, N. G. Tetrahedron Lett. 1986, 27, 877-890.
13. For a related example, see: Harnden, M. R.; Jarvest, R. L.; Bacon, T. H.; Boyd, M. R. J. Med. Chem. 1987, 30, 1636-1642.
14. Kjellberg, J.; Johansson, N. G. Tetrahedron 1986, 42, 6541-6544.

15. Kim, C. U.; Luh, B. Y.; Misco, P. F.; Bronson, J.; Hitchcock, M. J. M.; Ghazzouli, I.; Martin, J. C. 8th International Round Table Meeting on Nucleosides, Nucleotides, and their Biological Applications, Orange Beach, AL, October 1988, Abstract No. 41.
16. Binder, J.; Zbiral, E. Tetrahedron Lett. 1986, 27, 5829-5832.
17. Balzarini, J.; Naesens, L.; Herdewijn, P.; Rosenberg, I.; Holy, A.; Pauwels, R.; Baba, M.; Johns, D. G.; De Clercq, E. Proc. Natl. Acad. Sci. USA 1989, 86, 332-336.
18. Lin, T-S.; Chen, M. S.; McLaren, C.; Gao, Y-S.; Ghazzouli, I.; Prusoff, W. H. J. Med. Chem. 1987, 30, 440-444.
19. Kelsey, D. K.; Kern, E. R.; Overall, J. C., Jr.; Glasgow, L. A. Antimicrob. Agents Chemother. 1976, 9, 458-464.

RECEIVED February 23, 1989

Chapter 6

Synthesis and Antiviral Activity of Nucleotide Analogues Bearing the (*S*)-(3-Hydroxy-2-phosphonylmethoxy)propyl Moiety Attached to Adenine, Guanine, and Cytosine

Joanne J. Bronson[1], Ismail Ghazzouli[1], Michael J. M. Hitchcock[1], Robert R. Webb II[1], Earl R. Kern[2], and John C. Martin[1]

[1]Bristol-Myers Company, Pharmaceutical Research and Development Division, 5 Research Parkway, Wallingford, CT 06492–7660
[2]University of Alabama, Birmingham, AL 35294

The nucleotide analogues (S)-9-((3-hydroxy-2-phosphonylmethoxy)propyl)adenine (HPMPA, **1**), (S)-9-((3-hydroxy-2-phosphonylmethoxy)propyl)guanine (HPMPG, **3**) and (S)-1-((3-hydroxy-2-phosphonylmethoxy)propyl)cytosine (HPMPC, **4**) have been synthesized on a multigram scale and evaluated in vitro and in vivo for antiviral efficacy against herpesviruses. HPMPC emerged as the most selective antiherpes agent of this series. It showed a much greater in vivo efficacy than acyclovir against herpes simplex virus types 1 and 2 and also was more potent than ganciclovir in a murine cytomegalovirus infection model.

Much of the research directed towards the discovery of compounds with antiviral activity has focused on the synthesis and evaluation of nucleoside analogues. In most cases, these analogues, such as acyclovir (ACV) and ganciclovir (DHPG), have a mechanism of action that involves enzymatic phosphorylations to first give a monophosphate which in turn is converted to a diphosphate. Ultimately, the triphosphate is formed which inhibits the virus specified DNA polymerase and thus virus replication. The phosphorylated species are called nucleotide analogues, and some stable nucleotide analogues, including phosphonate derivatives of ACV (1) and DHPG (2,3), have been reported to have antiviral activity.

The discovery by De Clercq and Holy of (S)-9-((3-hydroxy-2-phosphonylmethoxy)propyl)adenine (HPMPA, **1**) as a potent, broad spectrum antiviral agent has defined a new class of nucleotide analogues structurally characterized as phosphonylmethyl ethers of acyclic nucleoside derivatives (4). In that publication and in additional studies, HPMPA was found to be active in vitro against a variety of DNA viruses including herpes simplex virus types 1 and 2 (HSV 1 and 2), cytomegalovirus (CMV), varicella-zoster virus (VZV) (5), Epstein-Barr virus (6), adenovirus (7), and a retrovirus (Maloney sarcoma virus). In vivo activity against herpes simplex virus type 1 was demonstrated by the i.p. and topical routes of administration (4).

0097–6156/89/0401–0088$06.00/0

HPMPA (**1**) has been proposed to be an analogue of 2'-deoxyadenylic acid (**2**) and to have a mechanism of action that involves two phosphorylations to give a triphosphate analogue which then inhibits viral DNA polymerases (8). This mechanism is similar to that described above for acyclovir. The difference is that acyclovir must first be phosphorylated to the monophosphate; whereas, HPMPA is a monophosphate equivalent. Acyclovir derives selectivity at this first step in that the phosphorylation is only efficiently carried out by virus specified enzymes (9). However, the result is that acyclovir has a limited spectrum of action and is not active against viruses that lack a kinase for which it is a substrate. These viruses include CMV and thymidine kinase deficient mutants of herpes simplex virus. In bypassing this first step of activation, HPMPA is active against such viruses (4). In fact, HPMPA has been reported to have in vivo activity against a thymidine kinase deficient, acyclovir resistant strain of HSV 1 in a rabbit model of herpes keratitis (10).

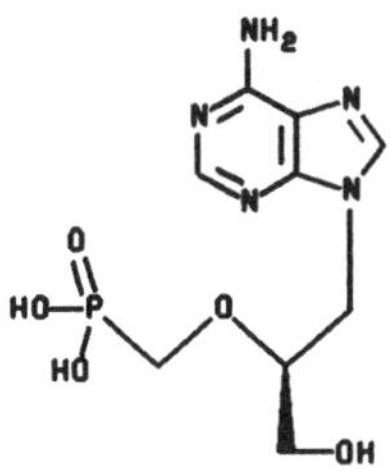

1 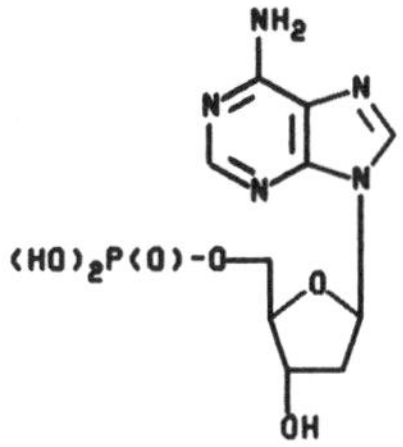 2

The structural analogy of HPMPA to 2'-deoxyadenylic acid (**2**) also provides a rationale for the fact that the S isomer as drawn is the active form of HPMPA (4). This S enantiomer has the same chirality as phosphate **2**. The opposite enantiomer of HPMPA, the R isomer, is inactive.

The guanine, cytidine, and thymidine analogues related to HPMPA have been described (5). HPMPG (**3**) and HPMPC (**4**) are active against DNA viruses, but HPMPT (**5**) is not. In this chapter, we describe the synthesis and antiviral comparison of the three active members of this series, HPMPA, HPMPG, and HPMPC.

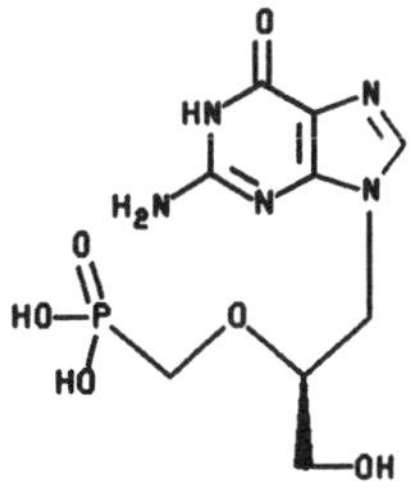

3 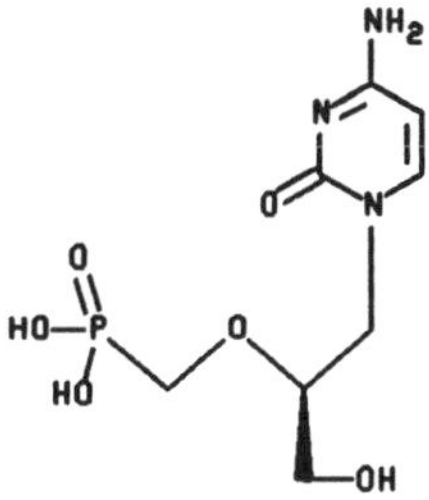 4 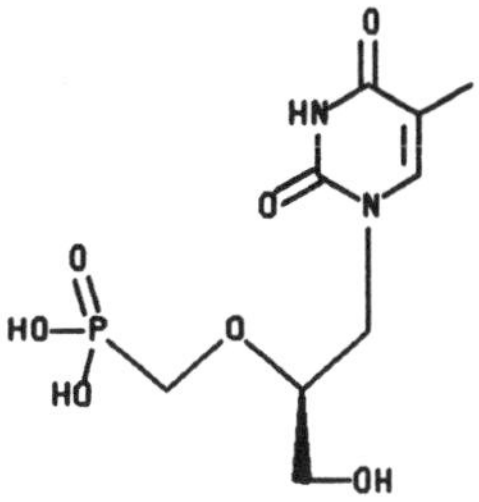 5

Chemistry

The preparation of compounds in this series of nucleotide analogues requires the recognition of two synthetic complications in addition to the well known problems associated with the coupling of nucleoside bases to side chains. One requirement is that only the active S enantiomer should be prepared: therefore, the starting material for the side chain must be chiral. A second issue is that the terminal hydroxyl on the side chain must be protected so that the phosphonylmethyl functionality may be introduced selectively on the secondary alcohol.

Two possible synthetic approaches to this series are represented in Scheme 1. In the retrosynthetic analysis labeled A, a preformed acyclic nucleoside **6**, with the terminal hydroxyl protected with P, is alkylated to introduce the phosphonylmethyl ether functionality. The second strategy (B) proceeds via an electrophilic side chain derivative **7**, which contains the phosphonate functionality and a leaving group X, to alkylate the heterocyclic base. Because early introduction of the base provides more crystalline intermediates, strategy A is best for the synthesis of a specific compound and was used for the preparation of HPMPA. Since the preformed electrophile can be used for the alkylation of many heterocycles, the second alternative is ideal for the preparation of a number of analogues and was used for the synthesis of HPMPG and HPMPC.

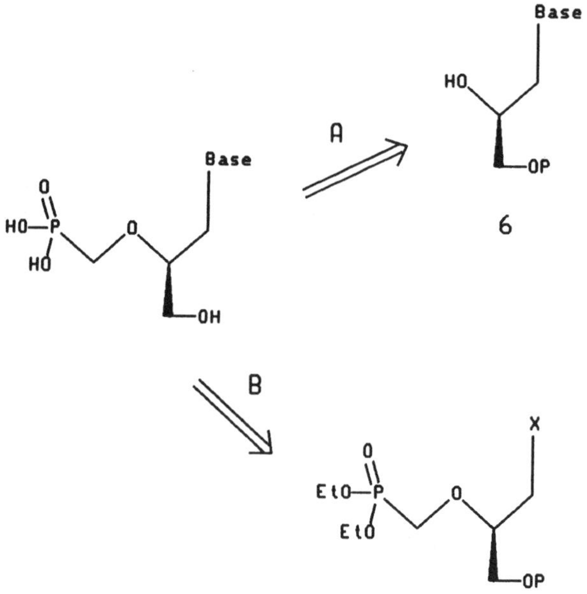

Scheme 1

HPMPA Synthesis. Our account of a multigram synthesis of HPMPA has appeared (11). The chiral starting material was (S)-9-(2,3-dihydroxypropyl)adenine (DHPA, **8**), which was readily synthesized from adenine and (S)-glycerol acetonide (12). Briefly, DHPA (**8**) was bis-protected to afford **9** in 65 % yield by reaction with trityl chloride and triethylamine in DMF (Scheme 2). The secondary hydroxy functionality of **9** was alkylated by successive treatment with NaH and then diethyl tosyloxymethylphosphonate (13) in DMF to give fully protected HPMPA **10** in 90 % yield. The trityl groups were removed by hydrolysis with 80 % aqueous acetic acid to furnish **11** (85 %), and then cleavage of the ester functionalities of **11** was accomplished with ten equivalents of trimethylsilyl bromide (14) in DMF at room temperature for 5 h to furnish HPMPA (90 %). The workup of this final reaction involved the removal of all volatiles by evaporation. Addition of water resulted in hydrolysis of the intermediate silyl esters, and then dilution with acetone caused precipitation of HPMPA as the crystalline zwitterion. This simple workup procedure was effective for all the nucleotide analogues described in this chapter and avoids tedious ion exchange or reverse phase purifications. This synthetic approach to HPMPA allowed for the efficient preparation of multigram quantities in 36 % overall yield from adenine.

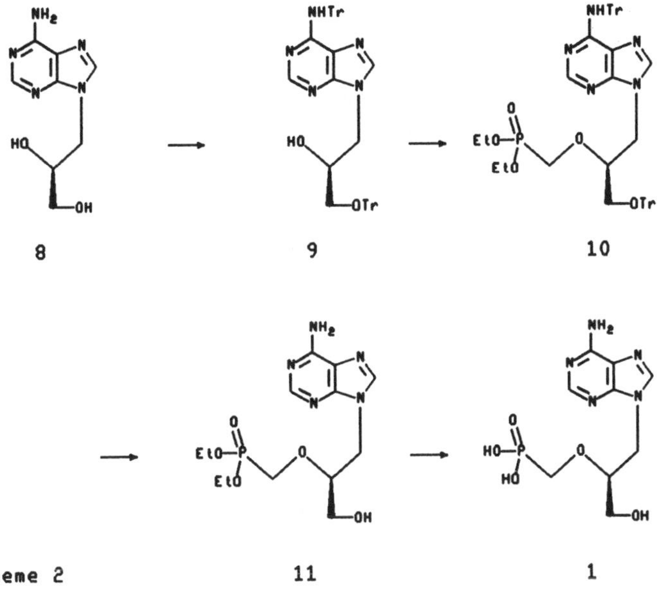

NH2 NHTr NHTr
HO OH OTr EtO P O EtO
8 9 10
11 1

Scheme 2

HPMPG Synthesis. HPMPG was prepared by the more general approach B. First, the appropriate side chain bearing a phosphonate functionality was prepared (Scheme 3). Phase transfer benzylation of R-glycerol acetonide (**12**) followed by acid hydrolysis by the literature procedure worked well on a large scale to give (S)-1-O-benzylglycerol (**13**) (15). Reaction of **13** with monomethoxytrityl chloride, 4-dimethylaminopyridine, and triethylamine in dichloromethane gave

14 in quantitative yield without purification. Treatment of **14** with NaH in THF at reflux and then diethyl tosyloxymethylphosphonate (13) (0 °C to room temperature, 14 h) furnished phosphonylmethyl ether **15** (55 % yield from **13**, chromatographed, 50 % to 75 % ethyl acetate/hexanes). Detritylation of **15** with 80 % aqueous acetic acid (steam bath, 20 min) or Amberlyst-15 in methanol (room temperature, 16 h) afforded **16** in 86 - 90 % yield after chromatography (75 % ethyl acetate/hexane to 8 % ethanol/ethyl acetate). Treatment of **16** with tosyl chloride in pyridine gave tosylate **17** in 91 % yield after chromatography. Alternatively, reaction of compound **16** with methanesulfonyl chloride and triethylamine in dichloromethane gave mesylate **18** in quantitative yield without purification. The ease of the synthesis of **18** makes it an intermediate superior to **17**.

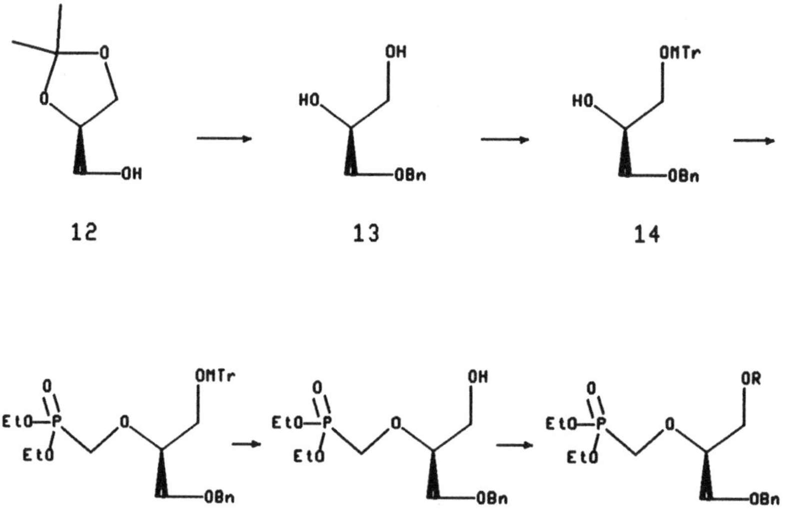

12 13 14

15 16 17, R=Ts; 18, R=Ms

Scheme 3

A solution of **17** in DMF at 80 °C was treated in one batch with a solid mixture of 2-amino-6-benzyloxypurine (16,17) and cesium carbonate to give after workup and chromatography (5 % methanol/ethyl acetate) a 59 % yield of **19** as an oil (Scheme 4). A more polar minor byproduct was identified as the N-7 isomer **20**, and the assignments of the structures of **19** and **20** were as based on NMR spectroscopic data (18). The first step of the deprotection of **19** was achieved by transfer hydrogenation (20 % $Pd(OH)_2/C$, cyclohexene, ethanol, reflux) to afford a 90 % yield of **21** as a foam (chromatographically purified with 10 % methanol/dichloromethane). Finally, treatment of diethyl ester **21** with a ten fold excess of bromotrimethylsilane in DMF at room temperature for 5 h gave HPMPG (**3**) in 68 % yield following recrystallization from ethanol/water.

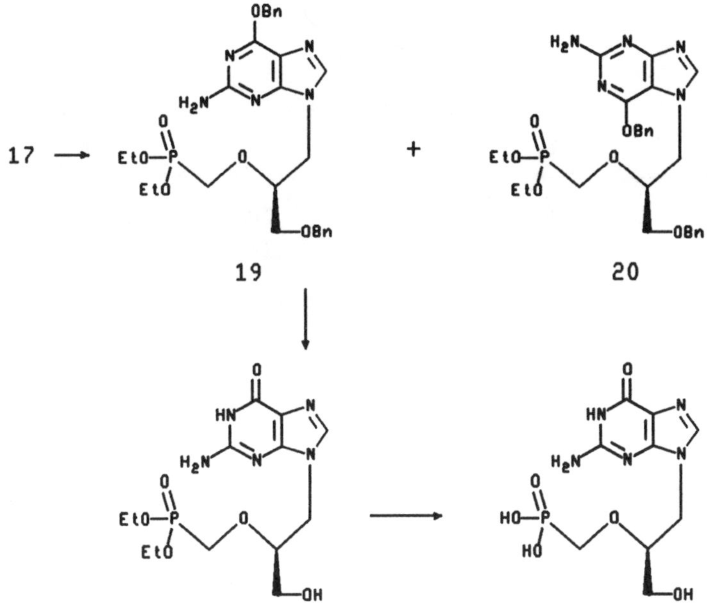

Scheme 4

HPMPC Synthesis. Mesylate **18** was utilized for the preparation of HPMPC (Scheme 5) (19). A mixture of **18** (24 mmol), cytosine (29 mmol) and cesium carbonate (49 mmol) in 50 mL of DMF was heated at 90 °C for 2.5 h. After evaporation, the crude product was purified by chromatography (5 - 10 % methanol/dichloromethane) to give the desired N-alkylated isomer **22** (66 %) and an O-alkylated byproduct **23** (23 %). The structural assignments of **22** and **23** were based on C-13 NMR spectroscopy in comparison to cytidine. The desired isomer **22** was next subjected to transfer hydrogenation (20 % palladium hydroxide on carbon, cyclohexene, ethanol, reflux for 4 h and then repeated for 8 h) to furnish the diethyl ester of HPMPC **24** as a foam in 70 % yield after chromatography (7.5 to 10 % methanol/dichloromethane). This reaction was carefully monitored by tlc to minimize the formation of the byproduct resulting from over reduction, a dihydrouracil derivative. Also, we found that the yield was maximized by replacing the catalyst with fresh material after partial conversion to product. Finally, the typical trimethylsilyl bromide deesterification afforded a 95 % yield of HPMPC (**4**) as a white powder after precipitation with ethanol from an aqueous solution.

For in vivo studies, highly soluble forms of these nucleotide analogues were needed. HPMPC and the other nucleotide analogues were converted to more soluble sodium salts. The substances were mixed with water (20 mL per gram) and treated with 1.00 N NaOH until the pH was adjusted to 7. The resulting solutions were then lyophilized to give the nucleotide analogues as white, highly soluble powders.

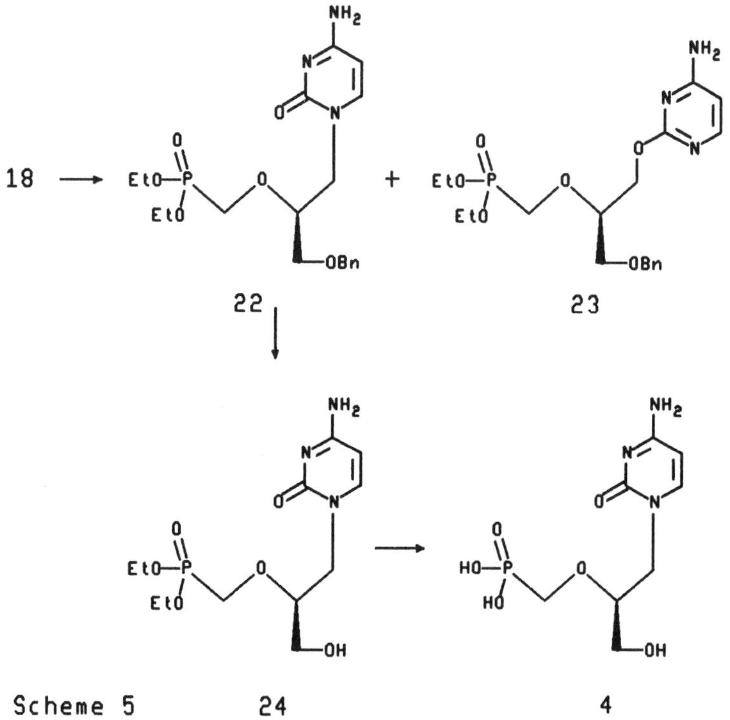

Scheme 5

In Vitro Data

The 50 % inhibitory doses of HPMPA, HPMPG and HPMPC in vitro against HSV 1 and 2 ranged from 2.3 to 25.2 μg/ml (Table I). These data closely parallel that reported by De Clercq and Holy (4) showing these substances to be approximately ten fold less potent than acyclovir (ACV) against HSV 1 and 2. However, the phosphonates are more potent than acyclovir against a thymidine kinase deficient strain of HSV 1, which is an indication that these nucleotide analogues do not rely on this kinase for activation.

Cytomegalovirus (CMV) is an opportunistic pathogen causing severe infections in many AIDS patients. Ganciclovir (DHPG) is an experimental therapy for infections caused by this virus (20). All three phosphonates exhibited in vitro potencies slightly superior to that of DHPG, and HPMPC was recently shown to have the best in vitro therapeutic index of this series (21). HPMPA and HPMPC are also active against murine cytomegalovirus, which is the virus used in animal model infection experiments in mice. In contrast to other phosphonates, these compounds with the HPMP side chain were not found to have in vitro activity against human immunodeficiency virus (22-24).

Table I. Antiviral Activity In Vitro

Virus	ID_{50}, μg/ml				
	HPMPA	HPMPG	HPMPC	ACV	DHPG
HSV 1 (BWS)	9.3	2.3	5.4	0.5	0.2
HSV 2 (G)	25.2	7.3	2.3	0.3	0.9
HSV 1 (TK-)	3.2	0.7	0.7	4.3	>10
HCMV (AD169)	0.27	0.39	0.22	--	1.8
MCMV (Smith)	0.02	--	0.02	--	0.06

In vitro antiviral data is sometimes not a reliable predictor of relative in vivo potencies. For instance, DHPG has an in vitro potency comparable to that of acyclovir against herpes simplex viruses in vitro but is one to two orders of magnitude more potent in a number of in vivo models (25,26). In order to assay for the true antiviral potential of these nucleotide analogues, extended in vivo studies were initiated.

Toxicity to Mice

To set doses for in vivo studies, the toxicities of the phosphonates were assessed in mice by oral (p.o.), intraperitoneal (i.p.), and intravenous (i.v.) routes (Table II). Three animals were assigned to each dose group. Compounds were given twice a day for five days except for the i.v. route where the animals were dosed once a day for only three days. HPMPG was highly toxic, especially by the i.p. route, killing all the mice even at the low dose of 10 mg/kg/day. HPMPA showed an intermediate toxicity resulting in death for all mice dosed at 100 mg/kg/day by the i.p. route but was not toxic at the dose level of 10 mg/kg/day. Presumably because of the fewer days of dosing, HPMPA and HPMPG were slightly less toxic when given by the i.v. rather than the i.p. route. HPMPC was by far the least toxic

Table II. Toxicity of HPMPA, HPMPG, and HPMPC to Mice

Dose, mg/kg/day	Route	% Survival		
		HPMPA	HPMPG	HPMPC
200	p.o.	100	33	100
100	p.o.	100	100	100
10	p.o.	100	100	100
200	i.p.	0	0	100
100	i.p.	0	0	100
10	i.p.	100	0	100
200	i.v.	0	0	100
100	i.v.	100	0	100
10	i.v.	100	100	100

and showed no effect on the mice dosed up to 200 mg/kg/day by all three routes.

HPMPA Pharmacokinetics in Mice

In order to evaluate the pharmacokinetics of a phosphonate analogue, a study of HPMPA in mice was undertaken. Each animal was dosed with 100 mg/kg of HPMPA. First, oral bioavailability was determined to be 5 % by a urinary recovery HPLC assay in which the percent of the dose recovered from five individual mice was averaged. Because of this low oral bioavailability, the initial efficacy evaluations of the nucleotide analogues were carried out by the i.p. or i.v. routes.

The plasma concentrations of HPMPA were measured as duplicate HPLC assays of plasma pooled from three mice. The data in Table III indicate that the overall exposure for the i.v. and i.p. routes of administration are similar which supports the use of the more easily performed i.p. dosing for model studies. A final point is that, after the one hour time point, the rate of clearance of HPMPA from the plasma slows dramatically indicating that this substance has a long terminal half-life. A possible explanation for this long terminal half-life is that the phosphonate nucleotide analogue does get into cells but is only slowly released. The slow release of the compound would lead to persistence in plasma which could result in an enhanced in vivo effect.

Table III. Pharmacokinetics of HPMPA in Mice

	Plasma Concentration of HPMPA, μg/ml		
Time, min	i.v.	i.p.	p.o.
5	196	117	0.40
15	63.4	110	0.63
30	25.5	52.6	0.63
45	10.6	27.5	0.78
60	4.54	8.65	0.86
90	3.45	5.76	0.56
120	3.94	5.11	--
150	3.50	3.32	--

Systemic Treatment of Herpes Simplex Virus Infections in Mice

HPMPA, HPMPG, and HPMPC were evaluated in mice for activity against herpes simplex virus types 1 and 2. These substances were compared to acyclovir in an encephalitis model, and efficacy is measured as % survival since untreated mice die of the infection. Animals were infected i.p. with the virus, and then treatment with a nucleotide analogue by various routes of administration was initiated 3 h post-infection. Animals were dosed with half the total daily dose indicated in the tables twice a day for 5 days. A * indicates a significance of <0.05 by the Fisher exact test.

HPMPA and HPMPG Against HSV. HPMPA and HPMPG were both efficacious in mouse mortality models of herpes simplex infections (Table IV), and these compounds did protect the mice in a dose dependent manner. HPMPG is the more potent analogue although the difference is somewhat exaggerated in Table IV because HPMPA was given by the less bioavailable oral route. An important point is that neither HPMPA or HPMPG provided complete protection. At higher doses both substances started to show toxicity before 100 % survival could be achieved. Thus, each compound had a similar but poor therapeutic index. The data for HPMPA is consistent with that reported by De Clercq and Holy who were able to achieve some, but not complete, protection in a systemic mouse model (4). The result of the higher potency and toxicity of HPMPG is similar to that recently described by Terry et al. for racemic HPMPG (27).

Table IV. In Vivo Efficacy of HPMPA (p.o.) Against HSV 2 and HPMPG (i.p.) Against HSV 1

	% Survival			
Dose, mg/kg/day	HPMPA, p.o.	ACV, p.o.	HPMPG, i.p.	ACV, i.p.
300	40*	90*	--	--
200	70*	67*	--	75*
100	59*	50*	--	70*
50	25*	34*	0	50*
25	17	--	--	--
10	--	--	10	--
1	--	--	60*	20
0.1	--	--	50*	--
0.05	--	--	30	--
Placebo	0	0	5	5

HPMPC Against HSV. The in vivo results against HSV with HPMPC provided the first indication that a phosphonylmethyl ether nucleotide analogue could have an improved therapeutic index over what was observed for HPMPA and HPMPG. In fact, HPMPC showed a toxicity substantially less than that of HPMPA and yet a potency comparable to that of the more active HPMPG. In the first in vivo study, the potency of HPMPC was underestimated since all the doses selected gave complete protection, Table V. All of the HPMPC treated mice survived an HSV 1 infection without evidence of drug toxicity when dosed by the i.p. route at 200, 100, and 10 mg/kg/day. Acyclovir was considerably less efficacious and gave at best a 75 % protection at the high dose of 200 mg/kg/day.

This model was repeated at lower doses to determine an end point for the activity of HPMPC. The data indicate that the 50 % effective dose of HPMPC was less than 1 mg/kg/day (Table VI). HPMPC was also given orally in this experiment and found to have a potency slightly superior to that of acyclovir.

Table V. In Vivo Efficacy of HPMPC (i.p.) Against HSV 1

Dose, mg/kg/day	% Survival HPMPC, i.p.	ACV, i.p.
200	100*	75*
100	100*	70*
10	100*	20
Placebo	5	5

Table VI. In Vivo Efficacy of HPMPC (i.p. and p.o.) Against HSV 1

	% Survival			
Dose, mg/kg/day	HPMPC, i.p.	ACV, i.p.	HPMPC, p.o.	ACV, p.o.
200	--	78*	70*	60*
100	--	--	90*	40
10	100*	--	0	23
1	90*	--	--	--
0.1	50 (p=0.08)	--	--	--
Placebo	16	16	16	16

The data obtained from the use of HPMPC against systemic infection with HSV 2 in mice (Table VII) were very similar to the results of the HSV 1 study. Significant protection against the type 2 virus was achieved at 0.1 mg/kg/day, and complete protection was found at the dose level of 10. The i.p. HPMPC data is a combination of two experiments. Although less potent by the oral route, HPMPC was again more efficacious than acyclovir.

Table VII. In Vivo Efficacy of HPMPC (i.p. and p.o.) Against HSV 2

	% Survival			
Dose, mg/kg/day	HPMPC, i.p.	ACV, i.p.	HPMPC, p.o.	ACV, p.o.
200	100*	100*	90*	50*
100	100*	20	70*	10
10	100*	10	30	0
1	60*	--	--	--
0.1	40*	--	--	--
Placebo	0	0	5	5

Topical Efficacy of HPMPC Against HSV 1

Acyclovir has shown a limited efficacy by the topical route in clinical trials on the treatment of primary genital herpes (28). HPMPC was compared to acyclovir in a cutaneous infection model in guinea pigs (29,30) (Table VIII). In this study, guinea pigs were infected on their backs in six regions and then treated with topical formulations in an attempt to reduce the severity of the lesions. Treatment was started 3 h post-infection and continued twice a day for 5 days. For HPMPC, the formulation was a tween 80/PEG vehicle. The commercial topical formulation of acyclovir was used. The lesions were scored each day for severity. A score of 0 indicates no lesion, and scores of 1 through 4 represent increasingly severe disease. The mean lesion score is a statistically derived representation of the overall severity of the disease (* indicates a significance of <0.01). The commercial 5 % formulation of acyclovir did not provide statistically significant therapeutic benefit. However, HPMPC was highly efficacious in a dose dependent manner and provided a substantial benefit even in a dilute formulation of 0.1 %.

Table VIII. Topical Efficacy of HPMPC Against HSV 1

Compound	Mean Lesion Score
5.0 % HPMPC	0.8*
1.0 % HPMPC	1.0*
0.1 % HPMPC	1.4*
5.0 % ACV	2.4
Virus Control	2.8
Placebo Control	3.0

Systemic Efficacy of HPMPC Against Murine Cytomegalovirus in Mice

No in vivo model is available for human cytomegalovirus, but the use of the mouse virus in mice is thought to be predictive for the human disease (3,31). In this mouse model, the efficacy of HPMPC and ganciclovir (DHPG) were compared (Table IX). Animals were infected with the virus by the i.p. route and then BID i.p. treatment was started at the indicated times of 6, 24, and 48 hours post-infection.

The data demonstrate that HPMPC is about ten fold more potent than DHPG in increasing the number of survivors at a given dose. For instance, HPMPC was protective at the low dose of 1.2 mg/kg/day; whereas, a significant effect with DHPG was achieved only at the 11.2 mg/kg/day dose. HPMPC was also efficacious when treatment initiation was delayed to 48 h after infection, which is a challenging regimen since at that time point untreated mice are quite sick and die on day 4 or 5. The apparent lack of efficacy for the highest doses of HPMPC and DHPG when treatment was initiated at 48 h post-infection was probably due to enhancement of the drug toxicity by the disease in these already sick animals.

Table IX. In Vivo Efficacy of HPMPC (i.p.) Against Murine Cytomegalovirus

Dose, mg/kg/day	Treatment Start, h	% Survival	
		HPMPC	DHPG
33.3	6	100*	93*
33.3	24	87*	67*
33.3	48	33	33
11.2	6	100*	67*
11.2	24	93*	80*
11.2	48	67*	73*
3.8	6	100*	53
3.8	24	53	13
3.8	48	67*	0
1.2	6	93*	7
1.2	24	60 (p=.06)	20
1.2	48	60 (p=.06)	20
Placebo	24	20	20
Control	--	27	27

Summary

Although possessing moderate in vitro potency relative to acyclovir, HPMPC emerged from this study as a highly efficacious antiviral agent in vivo against herpesviruses. The potency of HPMPC was dramatically demonstrated by its activity down to a low i.p. dose of only 0.1 mg/kg/day in mouse mortality models of HSV 1 and 2 infections. Also in mice, no evidence of toxicity was apparent at doses up to 200 mg/kg/day, which results in a very high therapeutic index for HPMPC. Probably because of low oral bioavailability, HPMPC was found to be less active by oral administration, but it was still more potent than acyclovir by this route. In addition, 0.1 % HPMPC was more efficacious than commercial 5 % acyclovir in a topical formulation against an HSV 1 cutaneous infection in guinea pigs.

In a murine cytomegalovirus infection model, HPMPC was shown to be significantly more potent than ganciclovir by providing protection down to a low dose of 1.2 mg/kg/day. Ganciclovir is currently under evaluation for the treatment of CMV infections in the clinic but has a limited therapeutic index due to myelosuppression (20). Especially since HPMPC was recently reported to be no more toxic than ganciclovir to human bone marrow cells (32), this nucleotide analogue merits further evaluation for possible therapy against cytomegalovirus infections in immunocompromised patients.

Literature Cited

1. Sidwell, R. W.; Huffman, J. H.; Barnard, D. L.; Reist, E. J. 8th International Round Table Meeting on Nucleosides, Nucleotides and Their Biological Applications Abstract 14, Orange Beach, AL, October 3, 1988.

2. Prisbe, E. J.; Martin, J. C.; McGee, D. P. C.; Barker, M. F.; Smee, D. F.; Duke, A. E.; Matthews, T. R.; Verheyden, J. P. H. J. Med. Chem. 1986, 29, 671-675.
3. Duke, A. E.; Smee, D. F.; Chernow, M.; Boehme, R.; Matthews, T. R. Antiviral Research 1986, 6, 299-308.
4. De Clercq, E.; Holy, A.; Rosenberg, I.; Sakuma, T.; Balzarini, J.; Maudgal, P. C. Nature 1986, 323, 464-467.
5. De Clercq, E.; Sakuma, T.; Baba, M.; Pauwels, R.; Balzarini, J.; Rosenberg, I.; Holy, A. Antiviral Research 1987, 8, 261-272.
6. Lin, J-C.; De Clercq, E.; Pagano, J. S. Antimicrob. Agents Chemother. 1987, 31, 1431-1433.
7. Baba, M.; Mori, S.; Shigeta, S.; De Clercq, E. Antimicrob. Agents Chemother. 1987, 31, 337-339.
8. Votruba, I.; Bernaerts, R.; Sakuma, T.; De Clercq, E.; Merta, A.; Rosenberg, I.; Holy, A. Mol. Pharmacol. 1987, 32, 524-529.
9. Elion, G. B.; Furman, P. A.; Fyfe, J. A.; de Miranda, P.; Beauchamp, L.; Schaeffer, H. J. Proc. Natl. Acad. Sci. USA 1977, 74, 5716-5720.
10. Maudgal, P. C.; De Clercq, E.; Huyghe, P. Invest. Ophthamol. Vis. Sci. 1987, 28, 243-248.
11. Webb, R. R.; Martin, J. C. Tetrahedron Lett. 1987, 28, 4963-4964.
12. Holy, A. Collect. Czech. Chem. Commun. 1975, 40, 187-214.
13. Holy, A.; Rosenberg, I. Collect. Czech. Chem. Commun. 1982, 47, 3447-3463.
14. McKenna, C. E.; Schmidhauser, J. J. Chem. Soc. Chem. Commun. 1979, 739.
15. Golding, B. T.; Ioannou, P. V. Synthesis 1977, 423-424.
16. MacCoss, M.; Chen, A.; Tolman, R. L. Tetrahedron Lett. 1985, 26, 1815-1818.
17. Robins, M. J.; Robins, R. K. J. Org. Chem. 1969, 34, 2160-2163.
18. Kjellberg, J.; Johansson, N. G. Tetrahedron 1986, 42, 6541-6544.
19. Webb, R. R.; Wos, J. A.; Bronson, J. J.; Martin, J. C. Tetrahedron Lett. 1988, 29, 5475-5478.
20. Collaborative DHPG Treatment Study Group, N. Engl. J. Med. 1986, 314, 801-805.
21. Snoeck, R.; Sakuma, T.; De Clercq, E.; Rosenberg, I.; Holy, A. Antimicrob. Agents Chemother. 1988, 32, 1839-1844.
22. Pauwels, R.; Balzarini, J.; Schols, D.; Baba, M.; Desmyter, J.; Rosenberg, I.; Holy, A.; De Clercq, E. Antimicrob. Agents Chemother. 1988, 32, 1025-1030.
23. Balzarini, J.; Naesens, L.; Herdewijn, P.; Rosenberg, I.; Holy, A.; Pauwels, R.; Baba, M.; Johns, D. G.; De Clercq, E. Proc. Natl. Acad. Sci. USA 1989, 86, 332-336.
24. Ghazzouli, I.; Bronson, J. J.; Russell, J. W.; Klunk, L. J.; Bartelli, C. A.; Franco, C.; Hitchcock, M. J. M.; Martin, J. C. J. Cell. Biochem. 1989, S13B, 300.
25. Martin, J. C.; Dvorak, C. A.; Smee, D. F.; Matthews, T. R.; Verheyden, J. P. H. J. Med. Chem. 1983, 26, 759-761.
26. Smee, D. F.; Martin, J. C.; Verheyden, J. P. H.; Matthews, T. R. Antimicrob. Agents Chemother. 1983, 23, 676-682.
27. Terry, B. J.; Mazina, K. E.; Tuomari, A. V.; Haffey, M. L.; Hagen, M.; Feldman, A.; Slusarchyk, W. A.; Young, M. G.; Zahler, R.; Field, A. K. Antiviral Research 1988, 10, 235-252.

28. Corey, L.; Nahmias, A. J.; Guinan, M. E.; Benedetti, J. K.; Critchlow, C. W.; Holmes, K. K. N. Engl. J. Med. 1982, 306, 1313-1319.
29. Mansuri, M. M.; Ghazzouli, I.; Chen, M. S.; Howell, H. G.; Brodfuehrer, P. R.; Benigni, D. A.; Martin, J. C. J. Med. Chem. 1987, 30, 867-871.
30. Alenius, S.; Oberg, B. Arch. Virol. 1978, 58, 277.
31. Kelsey, D. K.; Kern, E. R.; Overall, J. C., Jr.; Glasgow, L. A. Antimicrob. Agents Chemother. 1976, 9, 458-464.
32. Snoeck, R.; Lagneaux, L.; Bron, D.; De Clercq, E. 28th Interscience Conference on Antimicrobial Agents and Chemotherapy Abstract 739, Los Angeles, CA, October 25, 1988.

RECEIVED February 23, 1989

Chapter 7

Design of Inhibitors of Herpes Simplex Virus Thymidine Kinase

J. A. Martin, I. B. Duncan, and G. J. Thomas

Research Division, Roche Products Limited, P.O. Box 8, Welwyn Garden City, Hertfordshire AL7 3AY, England

The role of herpes simplex virus thymidine kinase in the pathogenesis of infection, its mechanism of action and the current status with regard to the design of inhibitors are reviewed. Examples of inhibitors based on substrate and product analogues are described, some of which are very potent and highly selective for the viral enzymes. One class of inhibitors with high *in vitro* potency and selectivity has shown a protective effect in mice under certain conditions and it is possible that compounds of this type may have potential as antiviral agents in man.

Herpes simplex virus (HSV) types 1 and 2 cause mild to severe disease in man, primarily orofacial and genital infections. A feature of these viruses is their ability to become latent in neuronal ganglia, from whence re-expression causes recurrent clinical episodes. Thus far, the design of antiviral agents against HSV infections has centered on compounds that inhibit major viral enzymes in the lytic cycle, whereas processes that control latency and reactivation have been relatively neglected targets.

The Role of Thymidine Kinase in Herpes Simplex Virus Infections

Under normal growth conditions, eukaryotic cells obtain the thymidine nucleotides required for DNA synthesis through a *de novo* pathway, in which thymidine monophosphate is synthesised from deoxyuridine monophosphate. Thus, thymidine kinase (TK) is not required for normal cell growth but the synthesis of a new species of TK by HSV is presumably necessary to accommodate the increased demand for thymidine triphosphate to fuel viral DNA synthesis. It is more than twenty-five years since the first report (1) establishing the presence of a virus-encoded TK in HSV-infected mouse fibroblasts. The relevance of this virus-induced enzyme to productive and latent infection both in cell culture and *in vivo* has been described (1-4).

0097–6156/89/0401–0103$06.00/0

Virus mutants have been isolated which do not produce a thymidine kinase (TK^-) but retain an unimpaired ability to replicate in cell culture (5). Since TK^- mutants have reduced pathogenicity in experimental animal models (6-11) the expression of TK is probably an important determinant of virulence *in vivo*. These findings together with identifiable differences between the viral enzyme and its cellular counterpart with regard to immunogenicity, molecular weight, thermostability and particularly substrate specificity (12-16) make this function an attractive target for antiviral chemotherapy. Whilst it has been suggested (17,18) that specific inhibitors of viral TK may suppress one or more aspects of HSV disease complex, there have been remarkably few reported attempts to develop potent and specific inhibitors of this viral enzyme. Several studies have described compounds that inhibit TK from *Escherichia coli* (19), Walker 256 carcinoma (20,21), Yoshida sarcoma (22,23) and L1210 cells (24). Differences have been identified between the bacterial and mammalian enzymes but surprisingly these studies (25) did not include the viral enzyme.

Mechanism of Action

Thymidine kinase catalyses the phosphorylation of thymidine in the presence of divalent cations such as magnesium with adenosine triphosphate (ATP) as the conventional phosphate donor. The phosphoryl transfer occurs with inversion of configuration at phosphorus, which has been explained (26) by a single in-line group transfer between ATP and thymidine within the enzyme complex. A more extensive system has been described (27) in which ADP or AMP may also act as the phosphate donor in 5'-phosphotransferase reactions to thymidine and in support of this there is evidence (28) that HSV-1 TK is translated as three polypeptides that may exist as a heteromultimer with a potential multiplicity of substrate binding sites. Earlier studies (29-31) had been reported in which phosphorylated species of thymidine affected TK catalysed reactions but this work used crude enzyme preparations and the presence of interfering components cannot be ruled out. However, now that the nucleotide and primary amino acid sequences of viral TK are known (32-34), the enzyme has been cloned and expressed in prokaryotes (35,36), and directed mutagenesis studies have begun (37,38), there is the prospect that a full structural analysis of the enzyme may be available soon. Evidence could then be obtained to explain these earlier observations and enable the design of better inhibitors from molecular modelling studies. Thus far, the design of inhibitors has been based instead on either an understanding of the enzyme-catalysed chemical transformations that occur or on the biochemical properties of the enzyme itself. Future synthetic programmes could be based on substrate analogues, product analogues, metal chelation or allosteric inhibition.

Substrate Analogues as Inhibitors

Whilst the mammalian TK has a very stringent substrate specificity the viral enzyme is considerably more tolerant of nucleoside analogues (39). More precisely, the cellular enzyme can process only thymidine and a few close analogues, whereas the viral enzyme has an ability to phosphorylate a wide range of pyrimidines and even some purine nucleosides. This marked difference has been exploited by many research groups in the search for new antiviral agents, resulting notably in the discovery of acyclovir (40-42). Such compounds depend on selective activation by viral TK and cellular kinases to their triphosphates, which then inhibit the viral DNA polymerase. Clearly the mode of action of such antiviral nucleosides is not inhibition of TK *per se*, although they do compete with the natural substrate, thymidine, for phosphorylation and will therefore deplete the pool of thymidine nucleotides in infected cells.

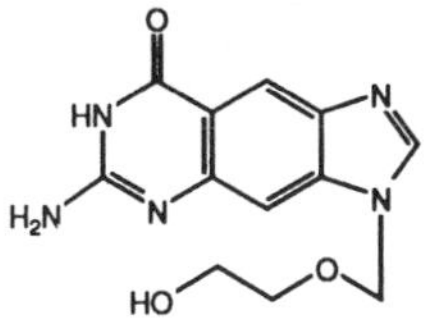

1 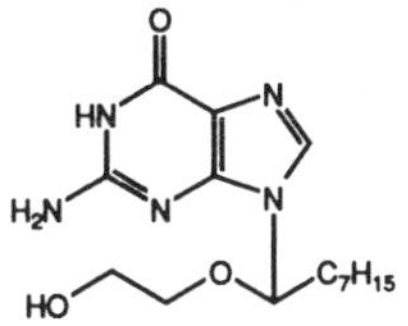 **2**

Analogues of acyclovir have been synthesised with modifications in the guanine and side chain moieties (43,44). These compounds had either weak or no activity against HSV-1 in a plaque reduction assay. In an attempt to rationalise the lack of antiviral activity the acceptability of these compounds as substrates for viral TK was measured. As might have been expected none of the compounds, including the acyclic analogues **1** and **2**, were phosphorylated. Both **1** and **2** inhibited the phosphorylation of acyclovir, indicating that they did bind to the enzyme. The development of more potent and selective inhibitors based on these lead structures has not been reported.

A series of N^2-phenyl substituted guanine derivatives was screened against HSV-1 TK in order to identify lead structures that could be modified to give potent and selective inhibitors (45,46). From the range of guanines screened (Table I) it is clear that substitution in the N^2-phenyl moiety has a marked effect on potency; for example, an electron-withdrawing group in either the meta or para position enhanced potency, whereas alkyl groups such as methyl or butyl in the para position reduced activity. However, a medium-sized alkyl residue such as ethyl was tolerated in the meta position. Glycosylation of the unsubstituted guanine **3** gave the corresponding 2'-deoxyguanosine derivative **9** which was the most active compound reported. Detailed studies have shown **9** to be selective for the isolated viral enzyme and it also inhibited TK in infected cells as measured by the entrapment of radiolabelled thymidine as nucleotides. At high concentration and with prolonged exposure **9** did exert some toxicity against the mammalian host cells but the leads generated in this study may provide a starting point for the development of more potent and selective inhibitors.

Wigdahl and Parkhurst (47) have shown 5-trifluorothymidine to be a competitive inhibitor of thymidine phosphorylation by both cellular and more effectively HSV-1 TK. It has also been proposed that the triphosphate of this compound might mimic thymidine triphosphate feedback regulation of both enzymes. It has been

Table I. Inhibition of HSV-1 Thymidine Kinase by N^2-Phenylguanine Derivatives

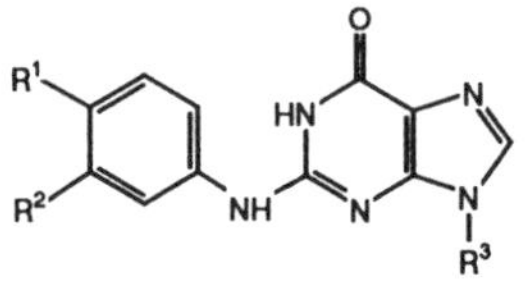

Compound	R^1	R^2	R^3	IC_{50} [μM] HSV-1
3	H	H	H	8
4	Me	H	H	50
5	n-Bu	H	H	50
6	H	Et	H	3
7	Br	H	H	1
8	H	Cl	H	1.3
9	H	H	2-deoxy-ß-D-ribofuranosyl	0.3

shown (48) that 5-trifluorothymidine is a substrate for both viral and host cell TK, and also that 5'-amino-5'-deoxythymidine selectively inhibits its phosphorylation by the cellular rather than the viral enzyme, so it is possible that closely related analogues may selectively inhibit the viral enzyme.

Product Analogues as Inhibitors

Derivatives of 2'-Deoxy-5-iodouridine. In an attempt to improve the antiviral efficacy of 5'-amino-2',5'-dideoxy-5-iodouridine a series of N-acyl derivatives was prepared as potential prodrugs (49). Whilst none of the compounds showed antiviral activity in tissue culture, some were found to be inhibitors of HSV-1 TK (Table II). Recently, additional data have been reported (50) which include activity against the type 2 enzyme.

All of the compounds shown in Table II were more active against HSV-2 TK than against the type 1 enzyme which may be a property of the 5-iodo substituent as was observed with the parent 5'-amino nucleoside. Inhibitory activity against both viral enzymes appeared to correlate reasonably well with changes in lipophilicity arising from modifications in the 5'-substituent. The most active compound in the series, **14**, contained the valeryl side chain and was approximately twenty times more active against the type 2 enzyme. Interestingly, the separation in activity between the two viral enzymes increased with increasing overall potency of these inhibitors. The methanesulphonamide **16** was approximately twice as active as the corresponding acetamide **10** against HSV-2 TK but from this single example it is impossible to predict whether sulphonamides are in general more potent than the corresponding amides. In all cases acylation of the 3'-hydroxyl function markedly reduced potency. None of the amides in Table II showed antiviral activity in cell culture and these compounds were inactive against an HSV-1 infection _in vivo_.

Table II. Inhibition of Thymidine Kinase by N-Acyl Derivatives of 5'-Amino-2',5'-dideoxy-5-iodouridine

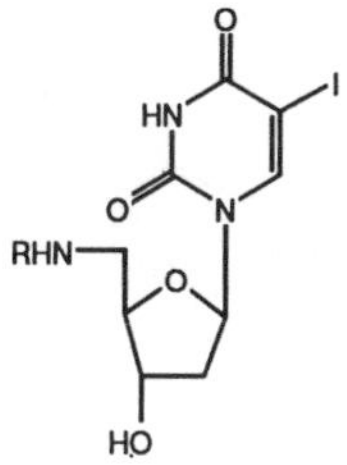

Compound	R	IC_{50} [μM]	
		HSV-1	HSV-2
10	CH_3CO	>200	78
11	C_2H_5CO	159	21
12	$(CH_3)_2CHCO$	55	7
13	$(CH_3)_2CHCH_2CO$	68	8
14	$CH_3(CH_2)_3CO$	38	2
15	PhCO	85	16
16	CH_3SO_2	>200	35

The 5'-azido derivative **17** has been reported (_49_) to be an inhibitor of HSV-1 TK (IC_{50} 280 μM). Although **17** was antiviral _in vivo_ it was not active in cell culture. Mechanism of action studies have not been reported, and so there is no evidence to associate the observed antiviral activity with inhibition of TK.

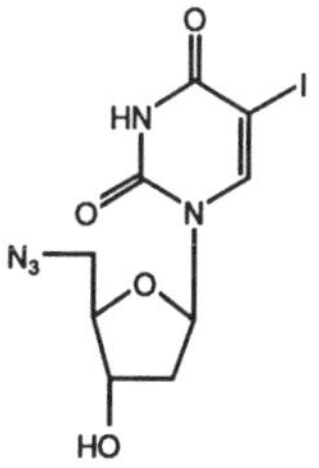

17

Derivatives of Thymidine. A series of sulphonamide derivatives of 5'-amino-5'-deoxythymidine (Table III) have been shown to inhibit

HSV TK (50) and have been compared with the amide derivatives described in the previous section.

Table III. Inhibition of Thymidine Kinase by Sulphonamide Derivatives of 5'-amino-5'-deoxythymidine

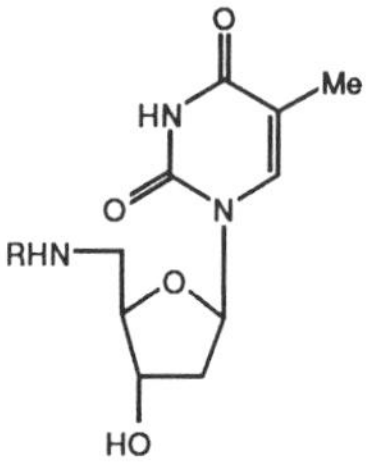

Compound	R	IC_{50} [μM] HSV-1	IC_{50} [μM] HSV-2
18	p-$CH_3C_6H_4SO_2$	>200	88
19	p-$CH_3OC_6H_4SO_2$	>200	116
20	p-$NO_2C_6H_4SO_2$	127	21
21	p-$BrC_6H_4SO_2$	171	13
22	CF_3SO_2	>200	>200
23	p-$HOSO_2C_6H_4SO_2$	>200	>200
24	$HOSO_2(CH_2)_3SO_2$	>200	>200

In common with amide derivatives (Table II) all of the active sulphonamides were significantly more potent against the type-2 enzyme. In the case of the benzenesulphonamides an electron-withdrawing group in the benzene ring markedly enhanced the potency, particularly against the type-2 enzyme. Compounds containing an acidic moiety in the 5'-side chain were totally inactive against both enzymes. An evaluation of the antiviral efficacy of these compounds has not been reported.

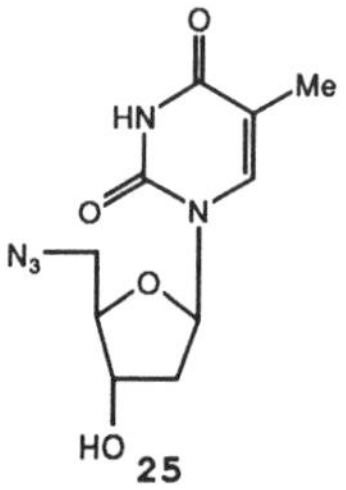

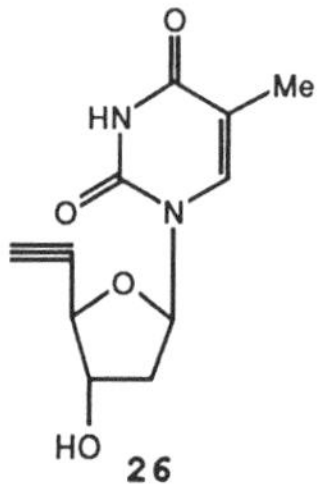

It is possible that 5'-ethynylthymidine **26** (51) was developed from the inhibitors **17** and **25** (49), and it is noteworthy that **26** has a totally different spectrum of activity compared with the

inhibitors described thus far, being more potent against the type 1 (K_i 0.09 μM) than the type 2 enzyme (K_i 0.38 μM). Since **26** did not inhibit human cytosolic TK, it must also be seen as highly selective for the viral enzymes. It was not cytotoxic in cell culture and did not inhibit host cellular DNA synthesis. Thymidine kinases with altered substrate specificity isolated from bromovinyldeoxyuridine (BVDU) and acyclovir-resistant virus strains were also inhibited by this compound. Mechanism of action studies in cell culture showed that the pool size of thymidine triphosphate was reduced but not the corresponding pools of triphosphates derived from adenosine, guanosine and cytosine respectively. As expected compound **26** was not antiviral *in vitro*, in agreement with the observation that viral TK is not essential for HSV replication in cell culture. However, **26** did reverse the antiviral effect of acyclovir, DHPG, FIAC, BVDU and 5'-amino-5'-deoxythymidine, all of which require an initial phosphorylation by viral TK for expression of activity. The evaluation of **26** *in vivo* has not been reported so far.

Derivatives of 2'-Deoxy-5-ethyluridine. A rational approach to the design of potent and selective inhibitors of HSV TK has been described recently (52-56). The dissociation constants of a series of 5-substituted uridine analogues against HSV and cellular TK (57) were examined (Table IV) in order to identify the nucleoside moiety most likely to confer selectivity for the viral enzyme.

Table IV. Dissociation Constants of Nucleoside Derivatives

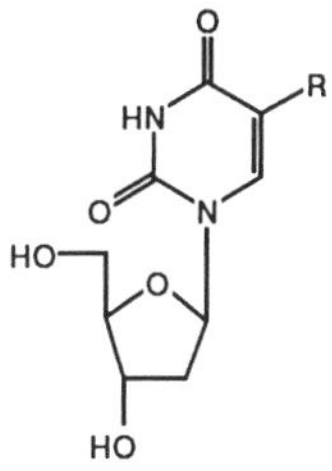

Compound	R	Dissociation Constant [μM]			
		Viral TK		Cellular TK	
		HSV-1	HSV-2	Cytosol	Mitochondria
27	I	0.6	0.3	7.4	8.2
28	CF_3	0.4	0.5	4.2	30
29	C_2H_5	0.7	0.3	82	30
30	n-C_3H_7	0.6	0.7	21	18
31	n-C_4H_9	1.6	4.0	100	40
32	$CH{=}CH_2$	0.5	0.5	35	1.7
33	CH=CHBr	0.4	3.0	>100	0.9

The high affinity of idoxuridine **27** and trifluorothymidine **28** for the cellular cytosolic enzyme and the strong affinity of vinyldeoxyuridine **32** and bromovinyldeoxyuridine **33** for the mitochondrial enzyme was indicative of poor selectivity. The ethyl, propyl and butyl derivatives, **29**, **30** and **31** respectively, were more selective for the viral enzymes and it was reasoned that inhibitors based on **29** could be expected to show the highest potency and selectivity for HSV TK.

Table V. Product Analogues

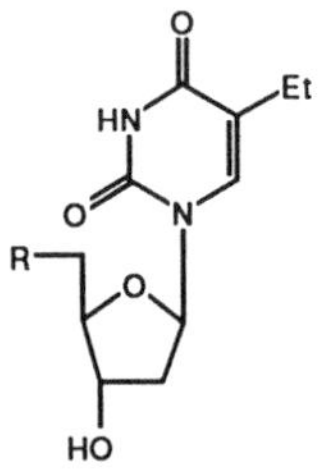

Compound	R	IC_{50} [μM] HSV-1	IC_{50} [μM] HSV-2
34	i-PrOP(O)(OH)O	208	40
35	MeP(O)(OH)O	320	-
36	PhP(O)(OH)O	72.6	20.3
37	$MeSO_2O$	8.1	4.8
38	$p\text{-}MeC_6H_4SO_2O$	12.7	4.1
39	$MeSO_2NH$	6.0	7.5
40	$PhSO_2NH$	15.2	4.6
41	MeCONH	15.8	4.7
42	PhCONH	3.1	3.2
43	$PhCH_2CONH$	1.0	0.3
44	$PhOCH_2CONH$	0.7	0.3

Having identified **29** as the preferred nucleoside moiety, derivatives containing functional groups isosteric and isoelectronic with the phosphate residue were prepared (Table V). The phosphate **34** and phosphonates **35** and **36** were rather poor inhibitors, whereas several other structural classes showed significantly better inhibition of both the HSV-1 and HSV-2 enzyme. In general, sulphonates and sulphonamides were more potent against the type 2 than the type 1 enzyme, while the benzamide **42** was identified as a potent inhibitor of both enzymes with scope for easy manipulation to afford analogues with enhanced potency. It was noted during these studies that homologation of the benzamide **42** increased potency, suggesting the presence of a hydrophobic interaction in the vicinity of the inhibitor binding site. In this series **43** and **44** were identified as good inhibitors, especially against the type 2 enzyme.

Having optimised the length of the linkage between the aryl residue and the nucleoside residue in **43**, additional analogues were prepared that contained spacer groups that were isosteric with the acetamide moiety. Whereas carbamate (OCONH, NHCOO) and urea (NHCONH) derivatives were less active, amine (CH_2CH_2NH), ketone (CH_2COCH_2) and alkane ($CH_2CH_2CH_2$) analogues had similar potency to the amide **43**, suggesting that the spacer group is not involved in interactions with the enzyme but serves only to locate the aryl residue in the optimum position.

A number of analogues of the phenylacetamide **43** which were substituted either in the phenyl ring, or on the α-carbon atom were prepared. It was found that a medium sized substituent in the ortho position of the phenyl ring increased potency by as much as ten-fold, as did a methyl or ethyl group on the α-carbon. In contrast, substitution on the α-carbon with polar groups such as hydroxyl or amino led to decreased activity. When these studies were extended to poly-substituted derivatives it was found that 2,6-disubstituted analogues were particularly active, the 2,6-dichloro and 2,6-dimethyl derivatives, **45** and **46**, had IC_{50} values of 0.003 μM and 0.008 μM respectively against the type 2 enzyme. As had been observed with the phenylacetamide **43** the corresponding amine, ketone and alkane derivatives also showed enhanced potency when 2,6-dichloro or 2,6-dimethyl substitution was introduced into the phenyl ring. In fact, compound **47** is one of the most potent inhibitors of HSV-2 TK known, with an IC_{50} of 0.0024 μM.

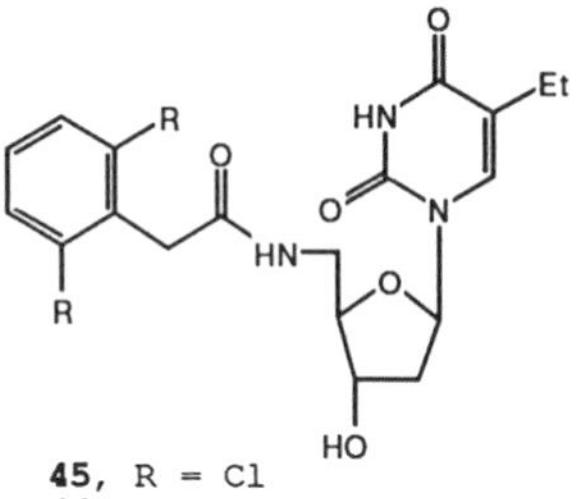

45, R = Cl
46, R = Me

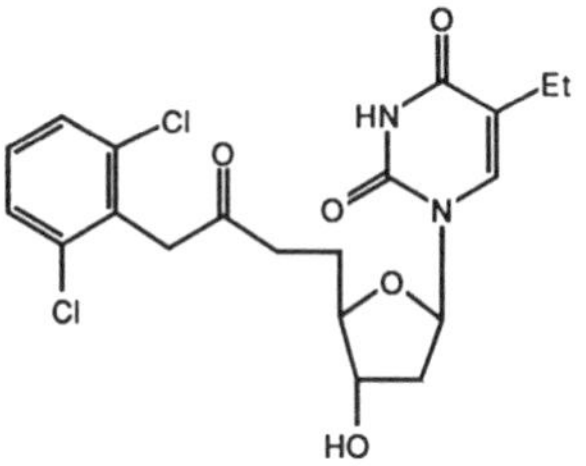

47

A similar series of substituted analogues of the phenoxyacetamide **44** was also studied. As in the case of phenylacetamides, substitution in the side chain by an alkyl group gave an increase in potency of almost twenty-fold. In contrast to the phenylacetamides, however, optimal activity was observed with a 2,4-disubstituted phenyl residue.

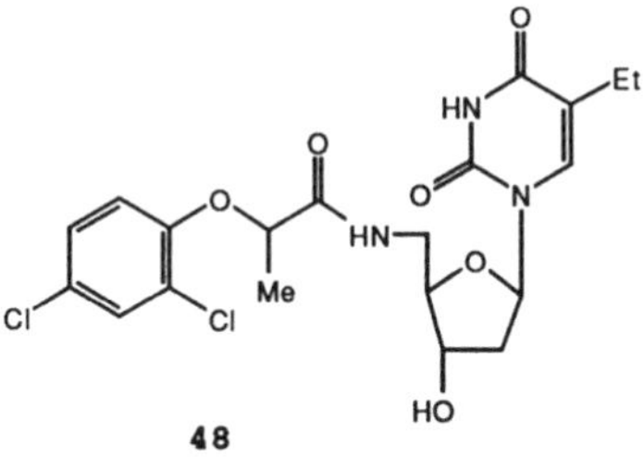

48

Replacement of the ether linkage by a sulphide, sulphoxide, sulphone or methylene group gave analogues with significantly reduced activity. One of the more potent compounds in the phenoxyacetamide series was **48** with an IC_{50} of 0.004 μM against the type 2 enzyme.

The potent inhibitors **45**, **47** and **48** were evaluated against cytoplasmic TK derived from two mammalian cell lines (Table VI), and a high degree of selectivity for the viral enzyme was observed in each case.

Table VI. Inhibition of Viral and Cellular Thymidine Kinase

Compound	IC_{50} [μM]			Selectivity
	HSV-2	Hela	Vero	
45	0.003	>200	>200	>60,000
47	0.0024	>200	>200	>83,000
48	0.004	>200	>200	>54,000

A study of the kinetics of inhibition of HSV-2 TK by the amide **48** showed it to be competitive with respect to thymidine, and non-competitive with respect to ATP. It was concluded that this compound located, as expected, at the thymidine binding site of the enzyme, and that the additional binding afforded by the 2,4-dichlorophenoxypropionamide residue, which contributed to the marked potency of this inhibitor, did not involve the ATP binding site.

As expected, none of these potent inhibitors showed antiviral activity in tissue culture, but compound **48** did show a marked antagonism of the antiviral activity of acyclovir in a plaque reduction assay, which probably resulted from inhibition of intracellular viral TK. Thymidine, at the same concentration, exhibited a similar antagonism of the antiviral effect of acyclovir. Compound **48** did produce a protective effect in mice infected with HSV-2, but the effect was variable and particularly sensitive to the strain of mouse, size of virus inoculum, formulation of test compound and route of administration. This _in vivo_ antiviral activity has not as yet been conclusively ascribed to inhibition of viral TK.

Conclusions

It is evident from data presented in this review that a number of research groups have prepared potent and selective inhibitors of HSV TK. Thus far, inhibitors have been designed and developed from either substrate or product analogues but alternative approaches could involve metal chelation, allosteric inhibition or bisubstrate mechanisms. Indeed, the bisubstrate approach (_58,59_) has been successful in the design of inhibitors of bacterial (_60_) and mammalian deoxynucleoside kinases (_61-63_). Inhibitors of HSV TK that are available now may prove to be useful biochemical tools to probe the structure and function of the active site of the viral enzymes. Improved knowledge of the enzyme active site through further biochemistry and molecular biology should provide a better understanding of the molecular interactions that occur with both substrates and inhibitors. Meanwhile, the current generation of inhibitors may provide a means of studying the role of TK in the development of HSV infection in animal models, and may find a place in the management of herpes infections in man.

Literature Cited

1. Kit, S.; Dubbs, D. R. *Biochem. Biophys. Res. Comm.* **1963**, 11, 55.
2. Klemperer, H. G.; Hayes, G. R.; Shedden, W. J. H.; Watson, D. H. *Virology* **1967**, 31, 120.
3. Cheng, Y.-C. *Biochim. Biophys. Acta* **1976**, 31, 120.
4. Cheng, Y.-C.; Ostrander, M. J. *Biol. Chem.* **1976**, 251, 2605.
5. Dubbs, D. R.; Kit, S. *Virology* **1964**, 22, 493.
6. Field, H. J.; Wildy, P. J. *J. Hyg.* **1978**, 81, 267.
7. Tenser, R. B.; Miller, R. L.; Rapp, F. *Science* **1979**, 205, 915.
8. Stanberry, L. R.; Kit, S.; Myers, M. G. *J. Virol.* **1985**, 55, 322.
9. Gordon, Y.; Gilden, D. H.; Shtram, Y.; Asher, Y.; Tabor, E.; Wellish, M.; Snipper, D.; Hadar, J.; Becker, Y. *Arch. Virol.* **1983**, 76, 39.
10. Field, H. J.; Darby, G. *Antimicrob. Agents Chemother.* **1980**, 17, 209.
11. Tenser, R. B.; Ressel, S.; Dunstan, M. E. *Virology* **1981**, 112, 328.
12. Jamieson, A. T.; Gentry, G. A.; Subak-Sharpe, J. H. *J. Gen. Virol.* **1974**, 24, 465.
13. Thouless, M. E. *J. Gen. Virol.* **1972**, 17, 307.
14. Cheng, Y.-C.; Dutschman, G.; Fox, J. J.; Watanabe, K. A.; Machida, H. *Antimicrob. Agents Chemother.* **1981**, 20, (3), 420.
15. Kit, S.; Leung, W. C.; Jorgensen, G. N.; Dubbs, D. R. *Int. J. Cancer* **1974**, 14, 598.
16. Thouless, M. E.; Skinner, G. R. B. *J. Gen. Virol.* **1971**, 12, 195.
17. Sim, I. S.; McCullagh, K. G. In *Approaches to Antiviral Agents;* Michael R. Harnden., Ed.; Macmillan: London, **1985**; Chapter 2.
18. Cheng, Y.-C.; In *Antiviral Drugs and Interferon: The Molecular Basis of Their Activity;* Becker, Y., Ed.; Martinus Nijhoff Publishing: Boston, **1984**; Chapter 4.
19. Rohde, W. *FEBS Lett.* **1977**, 82, 118.
20. Baker, B. R.; Neenan, J. P. *J. Med. Chem.* **1972**, 15, 940.
21. Neenan, J. P.; Rohde, W. *J. Med. Chem.* **1973**, 16, 580.
22. Harrap, K. R.; Stringer, M.; Browman, G. P.; Dady, P. J.; Cobley, T. In *Advances in Tumour Prevention, Detection and Characterisation;* Davis, W., Harrap, K. R., Eds.; Excerpta Medica: Amsterdam, **1978**; Vol. 4, p.93.
23. Stringer, M. Ph.D. Thesis, University of London, **1974**.
24. Barrie, S. E.; Davies, L. C.; Stock, J. A.; Harrap, K. R. *J. Med. Chem.* **1984**, 27, 1044.

25. Hampton, A.; Kappler, F.; Chawla, R. R. *J. Med. Chem.* **1979**, 22, 1524.
26. Arnold, J. R. P.; Cheng, M. S.; Cullis, P. M.; Lowe, G. *J. Biol. Chem.* **1986**, 261, 1985.
27. Labenz, J.; Müller, W. E. G.; Falke, D. *Arch. Virol.* **1984**, 81, 205.
28. Marsden, H. S.; Haar, L.; Preston, C. M. *J. Virol.* **1983**, 46, 434.
29. Fyfe, J. A.; McKee, S. A.; Keller, P. M. *Mol. Pharmacol.* **1983**, 24, 316.
30. Labenz, J.; Friedrich, D.; Falke, D. *Arch. Virol.* **1982**, 71, 235.
31. Just, I.; Dundaroff, S.; Falke, D.; Wolf, H. U. *J. Gen. Virol.* **1975**, 29, 69.
32. Wagner, M. J.; Sharp, J. A.; Summers, W. C. *Proc. Nat. Acad. Sci. USA* **1981**, 78, 1441.
33. Swain, M. A.; Galloway, D. A. *J. Virol.* **1983**, 46, 1045.
34. Kit, S.; Kit, M.; Qavi, H.; Trkula, D.; Otsuka, H. *Biochem. Biophys. Acta* **1983**, 741 158.
35. Colbere-Garapin, F.; Chousterman, S.; Horodniceanu, F.; Kourilsky, P.; Garapin, A-C. *Proc. Nat. Acad. Sci. USA* **1979**, 76, 3755.
36. Waldman, A. S.; Haeusslein, E.; Milman, G. *J. Biol. Chem.* **1983**, 258, 11571.
37. Inglis, M. M.; Darby, G. *J. Gen. Virol.* **1987**, 68, 39.
38. Liu, Q. Y.; Summers, W. C. *Virology* **1988**, 163, 638.
39. Cheng, Y.-C. *Ann. N. Y. Acad. Sci.* **1977**, 284, 594.
40. Elion, G. B.; Furman, P. A.; Fyfe, J. A.; de Miranda, P.; Beauchamps, L.; Schaeffer, H. J. *Proc. Nat. Acad. Sci. USA* **1977**, 74, 5716.
41. Schaeffer, H. J.; Beauchamps, L.; de Miranda, P.; Elion, G. B.; Bauer, D. J.; Collins, P. *Nature, London* **1978**, 272, 583.
42. Fyfe, J. A.; Keller, P. M.; Furman, P. A.; Miller, R. L.; Elion, G. B. *J. Biol. Chem.* **1978**, 253, 8721.
43. Beauchamp, L. M.; Dolmatch, B. L.; Schaeffer, H. J.; Collins, P.; Bauer, D. J.; Keller, P. M.; Fyfe, J. A. *J. Med. Chem.* **1985**, 28, 982.
44. Keller, P. M.; Fyfe, J. A.; Beauchamp, L.; Lubbers, C. M.; Furman, P. A.; Schaeffer, H. J.; Elion, G. B. *Biochem. Pharmacol.* **1981**, 30, 3071.
45. Focher, F.; Hildebrand, C.; Freese, S.; Ciarrocchi, G.; Noonan, T.; Sangalli, S.; Brown, N.; Spadari, S.; Wright, G. *J. Med. Chem.* **1988**, 31, 1496.
46. Focher, F.; Sangalli, S.; Ciarrocchi, G.; Della Valle, G.; Talarico, D.; Rebuzzini, A.; Wright, G.; Brown, N.; Spadari, S. *Biochem. Pharmacol.* **1988**, 37, 1877.
47. Wigdahl, B. L.; Parkhurst, J. R. *Antimicrob. Agents Chemother.* **1978**, 14, 470.
48. Fischer, P. H.; Murphy, D. G.; Kawahara, R. *Mol. Pharmacol.* **1983**, 24, 90.
49. Markham, A. F.; Newton, C. R.; Porter, R. A.; Sim, I. S. *Antiviral Res.* **1982**, 2, 319.
50. Sim, I. S.; Picton, C.; Cosstick, R.; Jones, A. S.; Walker, R. T.; Chamiec, A. J. *Nucleosides and Nucleotides* **1988**, 7, 129.

51. Nutter, L. M.; Grill, S. P.; Dutschman, G. E.; Sharma, R. A.; Bobek, M.; Cheng, Y.-C. *Antimicrob. Agents Chemother.* **1987**, 31, 368.
52. Martin, J. A.; Duncan, I. B.; Hall, M. J.; Lambert, R. W.; Thomas, G. J.; Wong-Kai-In, P. *Second International Conference on Antiviral Research.* Williamsburg, Virginia, U. S. A., 10-14 April **1988**; *Antiviral Res.* **1988**, 9, 84.
53. Martin, J. A. *Second Symposium on Medicinal Chemistry in Eastern England.* Hatfield, Hertfordshire, England, 21 April **1988**.
54. Lambert, R. W.; Martin, J. A.; Thomas, G. J. E. P. 257378.
55. Lambert, R. W.; Martin, J. A.; Thomas, G. J. E. P. 256400.
56. Lambert, R. W.; Martin, J. A.; Thomas, G. J. E. P. 255894.
57. Cheng, Y.-C. In *Antimetabolites in Biochemistry, Biology and Medicine;* Skoda, J., Langen, P., Eds.; Pergamon Press: London, **1979**, p.263.
58. Wolfenden, R. *Accts. Chem. Res.* **1972**, 5, 10.
59. Lienhard, G. E. *Annu. Rep. Med. Chem.* **1972**, 7, 249.
60. Ikeda, S.; Chakravarty, R.; Ives, D. H. *J. Biol. Chem.* **1986** 261, 15836.
61. Bone, R.; Cheng, Y.-C.; Wolfenden, R. *J. Biol. Chem.* **1986**, 261, 5731.
62. Davies, L. C.; Stock, J. A.; Barrie, S. E.; Orr, M.; Harrap, K. R. *J. Med. Chem.* **1988**, 31, 1305.
63. Hampton, A.; Hai, T. T.; Kappler, F.; Chawla, R. R. *J. Med. Chem.* **1982**, 25, 801.

RECEIVED January 4, 1989

Chapter 8

Nucleotides as *Herpesvirus*-Specific Inhibitors of Protein Glycosylation

Roelf Datema[1] and Sigvard Olofsson[2]

[1]Bristol-Myers Company, Pharmaceutical Research and Development Division, 5 Research Parkway, Wallingford, CT 06492–7660
[2]Göteborgs Universitet, Department of Clinical Virology, Guldhedsgatan 10B, S–41346 Göteborg, Sweden

We describe a strategy to obtain herpesvirus-specific glycosylation inhibitors. This involves selective phosphorylation in infected cells of a nucleoside analog to a 5'-monophosphate, which inhibits translocation of sugar nucleotides from the cytoplasm into the Golgi-compartment where the terminal glycosyltransferases are located. Such an inhibitor may be useful in antiviral chemotherapy, or could serve as a tool to study the role of terminal glycosylation of viral glycoproteins in the intact host organism.

Viral proteins are N-glycosylated initially by the transfer *en bloc* of a glucosylated high-mannose type oligosaccharide from the lipid dolichol-diphosphate usually to a nascent protein (1). Transfer of non-glucosylated oligosaccharides has been observed (2), but, as yet, not with viral proteins. The N-linked oligosaccharides are processed by glycosidases, and some of this processing occurs in the rough endoplasmic reticulum (3). The extent of processing is dependent on a variety of conditions, which may include the location of a particular oligosaccharide on the protein (4), but the removal of glucose residues appears to be ubiquitous. The trimming of sugar residues and the subsequent addition of peripheral sugars, to result in so-called hybrid-type or complex-type oligosaccharides, occurs by concerted action of glycosidases and glycosyltransferases in the Golgi apparatus (9), and can result in a plethora of oligosaccharides. (See Figure 1 for one example.)

Some viral glycoproteins contain O-linked oligosaccharides, usually in addition to N-linked oligosaccharides (5). The O-linked oligosaccharides are assembled by stepwise addition of sugar residues onto a protein in transit through the Golgi apparatus (6). Thus, the synthesis of O-linked oligosaccharides occurs simultaneously with the addition of peripheral sugars to

0097–6156/89/0401–0116$06.00/0

N-linked oligosaccharides, and these processes are referred to as "terminal glycosylation" (7).

The biological role of terminal glycosylation of viral glycoproteins

Several low-molecular weight compounds (sugar analogs and alkaloids) are known to inhibit discrete steps of the trimming pathway, as shown in Figure 1, and have been used to study the biological role of the oligosaccharide processing using virus-infected cells in culture. It became rapidly apparent that the biological effects were strongly dependent on the particular protein or viral system (8). Examples in case are two retroviral systems. In both Rous Sarcoma Virus (RSV)- and murine leukemia virus (MuLV)-infected cells, blocking N-glycosylation interferes with correct proteolytic processing of the envelope glycoproteins to stable end-products and the incorporation of these proteins into virions (9-10). Allowing N-glycosylation but blocking oligosaccharide trimming by any of the glucosidase inhibitors of Figure 1 prevented the formation of complex-type oligosaccharides of the envelope glycoproteins, but did not prevent the formation of fully infections RSV-particles (11). In contrast, formation of infectious MuLV was inhibited by deoxynojirimycin, an inhibitor of glucosidase I (10).

Sindbis-virus infected cells presented an interesting system to study the role of processing of N-linked oligosaccharides. The E1 and E2 glycoproteins of Sindbis virus each contain two glycosylation sites (12), and one glycosylation site on E1 and E2 carries exclusively complex-type oligosaccharides when the virus is grown in avian or mammalian cells. Hsieh et al. (4) could show that the folding of the polypeptide chains determined the extent of processing. It is shown in Table 1 that when N-glycosylation is inhibited by tunicamycin, the precursor protein of E2 (pE2) is not cleaved to E2, the cell-surface expression of pE2 is inhibited, and virus budding decreases (13). Other work has shown that cleavage of pE2 to E2 is required for virus release (14). Allowing glycosylation to occur, but preventing glucose-trimming by the glucosidase I-inhibitors N-methyl deoxynojirimycin or castanospermine and thus equipping the viral glycoproteins with the oligosaccharides $Glc_3Man_{9,8}GlcNAc_2$, still prevented cleavage of pE_2 (and hence virus release), but allowed cell-surface expression of PE2 (15-16). The same result was obtained earlier with bromoconduritol, that caused equipping the viral glycoproteins with $GlcMan_{9,8}GlcNAc_2$ (17). Allowing the removal of the glucose group (but preventing mannose-trimming using deoxymannojirimycin), equipping the glycoprotein with $Man_{9,8}GlcNAc_2$-oligosaccharides, permitted cleavage of pE2 and allowed virus budding (16). Thus, the presence of a single glucose group per oligosaccharide chain determined whether or not the protein can be proteolytically cleaved. These and other studies (5) indicate an essential role of glucose trimming in the maturation of some viruses.

Sindbis virus cultured in the presence of the mannosidase-I

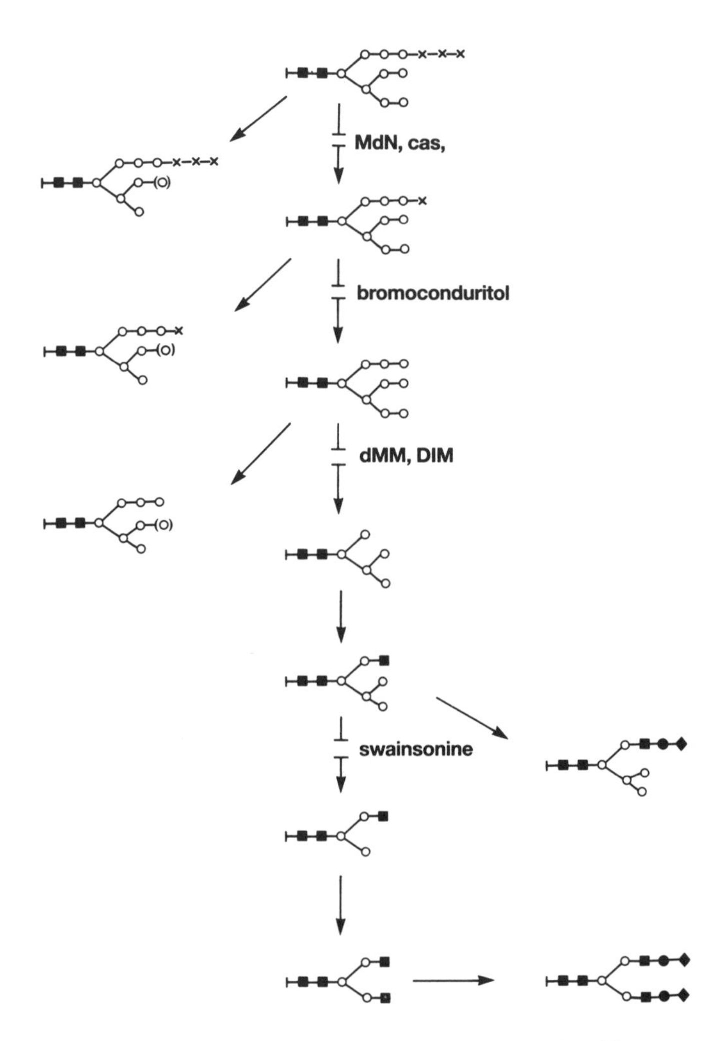

Figure 1. Processing of N-linked oligosaccharides. For abbreviations, see Table 1.
■, GlcNAc; O, Man; X, Glc; •, Gal; ♦, NeuAc.
DIM,1,4-dideoxy-1,4-imino-D-mannitol;
2,5-dihydroxymethyl-3,4-dihydroxypyrrolidine

inhibitor deoxymannojirimycin buds preferentially from internal membranes, rather than the cell-surface membrane (16). In the presence of the mannosidase-II inhibitor swainsonine (see Figure 1) the release of infectious Sindbis virus is not different from that of untreated cultures. In fact, the maturation of all enveloped viruses so far studied seems not to be affected by blocking oligosaccharide processing at the level of the Golgi enzymes N-acetylglucosaminyl transferase I or mannosidase II (5). Yet, many oligosaccharides of viral glycoproteins are processed past this stage to form complex-type oligosaccharides.

What, then, is the biological role of terminal glycosylation of viral glycoproteins? Based on evidence presented elsewhere, we suggested (5), that terminal glycosylation of viral glycoproteins may play a role at the virus-host level, for example in viral spread and pathogenesis. To study these phenomena in the intact organism, inhibitors active only in virus-infected cells are needed.

Inhibition of Terminal N- and O-Glycosylation Specific for Herpesvirus-Infected Cells

We chose to study the possibility of development of virus-specific inhibitors of glycosylation in the herpes simplex virus (HSV) system, because HSV has been widely used to study mechanisms of specific antiviral agents (18), and the pathogenesis of HSV has been studied in several animal models (19) allowing evaluation *in vivo*. The HSV-1 specified glycoprotein gC-1 is used as a molecular probe because terminal glycosylation of gC-1 may be important *in vivo*. Evidence for this are: removal of terminal sugars from gC-1 changes the antigenicity of the peptide part of the molecule (20), and the C3b-receptor activity of gC-1 is sialidase-dependent (21). Furthermore, gC-1 contains N- and O-linked oligosaccharides, which can be studied separately since tunicamycin-treatment does not jeopardize the proteolytic stability of gC-1 (22). Also, experimental tools to rapidly probe changes in N- and O-linked oligosaccharides of gC-1 have been developed (23-24).

Terminal glycosyltransferases are located inside the trans-Golgi compartment as is the acceptor glycoprotein (1). However, the substrates of the transferases, the sugar nucleotides are present in the cytosol. Hirschberg and co-workers (25) could show that sugar nucleotide translocator proteins in the Golgi membranes transport sugar nucleotides into the Golgi compartment and this translocation is coupled to export of the corresponding nucleoside 5'-monophosphate, as shown for UDP-Gal in Figure 2. The translocation of sugar nucleotides into the Golgi compartment can be inhibited by nucleoside 5'-monophosphates (26), and this phenomenon represents the target for the design of virus-specific glycosylation inhibitors. The inhibition of sugar nucleotide translocation limits the amount of the substrate for the glycosyltransferase, and causes a block in glycosylation. Herpesvirus specificity is obtained by selective phosphorylation of a nucleoside to an inhibitory nucleoside 5'-monophosphate in infected cells.

Table 1. Role of Glycosylation in the Sindbis Virus/BHK Cell System

Location of block	Inhibitor used	Cleavage $pE_2 \rightarrow E_2$	Cell-surface expression pE_2, E_2	Virus budding
LLO assembly	TM	No	No	Decreased
Glc trimming	Bc	No	Yes	Decreased
Glc trimming	dN, cas, MdN	No	Yes	Decreased
Man I trimming	dMM	Yes	Yes	Yes, but intracellularly
Man II trimming	swa	Yes	Yes	Yes

Abbreviations: LLO, lipid-linked oligosaccharide; TM, tunicamycin; Bc, bromoconduritol; dN, 1-deoxynojirimycin; cas, castanospermin; MdN, N-methyl-1-deoxynojirimycin; dMM, 1-deoxymannojirimycin; swa, swainsonin. For references, see text.

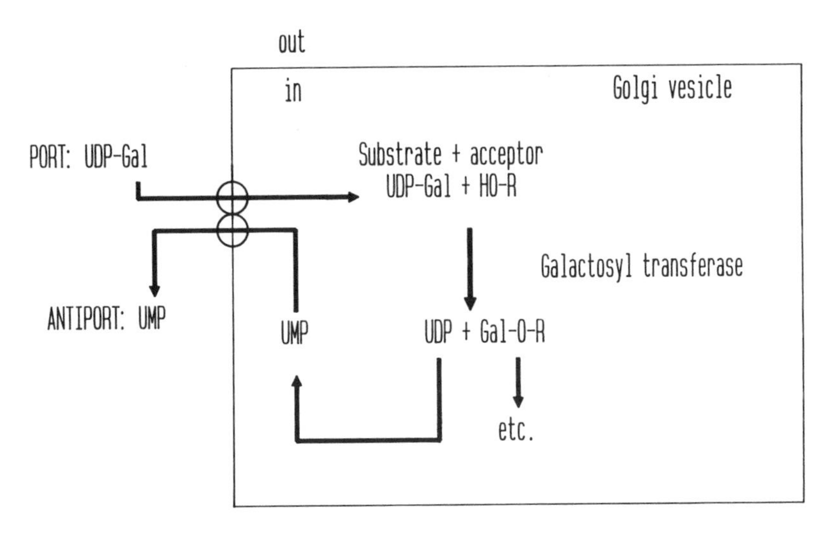

Figure 2. Schematic diagram of sugar nucleotide translocation.

We have studied (24, 27) the feasibility of this approach with the nucleoside analog (E)-5-(2-bromovinyl)-2'-deoxyuridine (BVdU). BVdU is a selective anti-HSV-1 agent interfering with viral DNA synthesis (18). The selectivity of BVdU is due in part to its phosphorylation by the HSV-induced thymidine kinase. In addition to interfering with viral DNA synthesis, the drug affects the synthesis of viral glycoproteins (24, 28-29), an effect also dependent on induction of the viral thymidine kinase. The effect on glycoproteins can be studied separately from the effect on DNA synthesis by adding the drug after the onset of viral DNA synthesis (i.e., 6 h p.i.). An analysis of the effect on glycoprotein gC-1 showed that incorporation of galactose and sialic acid into N-linked oligosaccharides and incorporation of sialic acid, and to a lesser extent, of galactose into O-linked oligosaccharides was blocked (27). This resulted in formation of normal amounts of gC-1, but with different oligosaccharides, i.e., with terminal GlcNAc and Fuc in N-linked oligosaccharides and terminal O-linked GalNAc.

This inhibition of terminal glycosylation was not caused by inhibition of formation of UDP-hexoses, inhibition of intracellular transport of glycoproteins in a galactosyl transferase-deficient cell line, nor by inhibition of acceptor glycoprotein synthesis. No evidence for formation of a sugar nucleotide analog of BVdU was obtained. The inhibition of terminal glycosylation did require formation of BVdU-5'-monophosphate, and was therefore not observed in cells infected with TK-negative variants of HSV-1. The block did occur also in HSV-2 infected cells, which can phosphorylate BVdU to its 5'-monophosphate (BVdUMP), but not any further. In a cell-free system BVdUMP inhibited the transport of pyrimidine sugar nucleotides (such as UDP-Gal, CMP-NeuAc) across Golgi membranes and, as a consequence, inhibited the terminal glycosylation. No inhibition of translocation of purine sugar nucleotide was observed, nor did BVdUMP inhibit galactosyl transferase.

Taken together, these results (27) show that it should be possible to obtain a nucleoside selectively phosphorylated by HSV thymidine kinase to a 5'-monophosphate which will inhibit sugar nucleotide translocation, in other words a virus-selective glycosylation inhibitor. Furthermore, it should be possible to obtain HSV-specific inhibitors of fucosylation or mannosylation by using guanosine analogs selectively phosphorylated by HSV-coded thymidine kinase (17, 30).

Implications

With the exception of chronic administration of oral acyclovir, chemotherapy has not been impressively useful in the management of recurrent HSV infections (31). Using therapeutic regimens of acyclovir, buciclovir or foscarnet effectively blocking HSV-1 replication and disease development in systemic, cutaneous and central nervous system-infection in mice, no clinical benefit was obtained when the virus (HSV-1) infected the skin of mice via sensory nerves (in zosteriform-spread models), despite the presence of immunity and drug availability (32). Thus, in these models of recrudescent disease, as in clinical trials of acute treatment of recurrent HSV infections, the benefits of potent and selective inhibitors of HSV replication were minimal. This implies that management of recurrent HSV-infections requires an approach other than inhibiting HSV DNA synthesis. Selective interference with viral protein glycosylation, resulting in altered immunogenicity of the viral glycoproteins may be such an approach.

Literature Cited

1. Kornfeld, R.; Kornfeld, S. Annu. Rev. Biochem. 1985, 54, 631-64.
2. Romero, P. A.; Herscovics, A. Carbohydr. Res. 1986, 151, 21-28.
3. Dunphy, W. G.; Rothman, J. E. Cell. 1985, 43, 13-21.
4. Hsieh, P.; Rosner, M. R.; Robbins, P. W. J. Biol. Chem. 1983, 258, 2555-61.
5. Datema, R.; Olofsson, S.; Romero, P. A. Pharmac. Ther. 1987, 33, 221-86.
6. Roseman, S. In Biology and Chemistry of Eucaryotic Cell Surfaces; Lee, E. Y. C.; Smith, E. E., Eds.; Academic: New York, 1974; pp. 317-54.
7. Schachter, H. Biol. Cell. 1984, 51, 133-46.
8. Schwarz, R. T.; Datema, R. Trends Biochem. Sci. 1984, 9, 32-34.
9. Schwarz, R. T.; Rohrschneider, J. M.; Schmidt, M. F. G. J. Virol. 1976, 12, 782-91.
10. Pinter, A.; Honnen, W. J.; Li, J. S. Virology. 1984, 136, 196-210.
11. Bosch, J. V.; Schwarz, R. T. Virology. 1984, 132, 95-109.
12. Strauss, E. G.; Strausse, J. H. In The Togavividae and Flaviviridae; Schlesinger, S.; Schlesinger, M. J., Eds.; Plenum Publishing Corp.: New York, 1986; pp. 35-90.
13. Schlesinger, M. J. In Virology; Fields, B. N.; Knipe, D. M.; Chancok, R. M.; Melnick, J. L. ; Roizman, B.; Shope, R. E., Eds.; Raven Press: New York, 1985; pp. 1021-32.
14. Garoff, J.; Kondor-Koch, C.; Riedel H. Curr. Top. Microbiol. Immunol. 1982, 99, 1-50.
15. Schlesinger, S.; Koyama, A. H.; Malfer, C.; Gee, S. L.; Schlesinger, M. J. Virus Res. 1985, 2, 139-49.
16. McDowell, W.; Romero, R. A.; Datema, R.; Schwarz, R. T. Virology. 1987, 161, 37-44.
17. Datema, R.; Romero, P. A.; Rott, R.; Schwarz, R. T. Arch. Virol. 1984, 81, 25-39.
18. De Clercq, E. Biochem. J. 1982, 2051, 1-13.
19. Stanberry, L. R. Progr. Med. Virol. 1986, 33, 61-77.
20. Sjöblom, I.; Lundström, M.; Sjögren-Jansson, E.; Glorisoso, J. C.; Jeansson, S.; Olofsson, S. J. Gen. Virol. 1987, 60, 545-54.
21. Smiley, M. L.; Friedman, H. M. J. Virol. 1985, 55, 857-61.
22. Svennerholm, B.; Olofsson, S.; Lundén, R.; Vahlne, A.; Lycke, E. J. Gen. Virol. 1982, 63, 343-49.

23. Lundström, M.; Olofsson, S.; Jeansson, S.; Lycke, E.; Datema, R.; Månsson, J.-E. Virology. 1987, 161, 385-94.
24. Olofsson, S.; Lundström, M.; Datema, R. Virology. 1985, 147, 201-05.
25. Hirschberg, C. B.; Snider, M. D. Annu-Rev. Biochem. 1987, 56, 63-88.
26. Capasso, J. M.; Hirschberg, C. B. Biochim. Biophys. 1984, 777, 133-239.
27. Olofsson, S.; Milla, M.; Hirschberg, C.; De Clercq, E.; Datema, R. Virology. 1988, 166, 440-450.
28. Misra, V.; Nelson, R. C.; Babiuk, L. A. Antimicrob. Agents Chemother. 1983, 23, 857-65.
29. Siegel, S. A.; Otto, M. J.; De Clercq, E.; Prusoff, W. H. Antimicrob. Agents Chemother. 1984, 25, 566-70.
30. Datema, R.; Ericson, A.-C.; Field, J. J.; Larsson, A.; Stenberg, K. Antiviral Res. 1987, 7, 303-16.
31. Douglas, J. M.; Critchlow, C.; Benedetti, J.; Mertz, G. J.; Connor, J. D.; Hintz, M. A.; Fahnlander, A.; Remington, M.; Winter, C; Corey, L. New Engl. J. Med. 1984, 310, 1551-56.
32. Kristofferson, A.; Ericson, A.-C.; Sohl-Åkerlund, A.; Datema, R. J. Gen. Virol. 1988, 69, 1157-66.

RECEIVED January 4, 1989

Chapter 9

Antiviral Activity of the Nucleotide Analogue Ribavirin 5′-Sulfamate

Donald F. Smee

Nucleic Acid Research Institute, 3300 Hyland Avenue, Costa Mesa, CA 92626

Ribavirin 5′-sulfamate inhibited Semliki Forest virus replication in cell culture and protected mice from a lethal virus challenge. This activity was unexpected since ribavirin was not inhibitory to the virus. Detailed mode of action studies showed that the analog inhibited viral polypeptide and RNA syntheses to a greater extent than comparable host cell functions. Viral RNA synthesis inhibition was caused by the inhibited synthesis of the viral RNA polymerase protein. The compound's primary mode of action was on aminoacyl-tRNA synthesis, the first step required for mRNA translation into protein. Other cellular processes affected as a result of protein synthesis inhibition included DNA and polyamine syntheses, and methylation and sulfuration reactions.

For years investigators have explored the potential of nucleotide analogs as inhibitors of viral infections (1), with much of the effort focused on modifications at the 5′-monophosphate position. To date, 5′-phosphonate derivatives of either naturally occurring nucleosides (2) or of nucleoside analogs (3) have not possessed the desired biological activity in cell culture. Part of the reason for this lack of activity is that these polar compounds do not readily penetrate into cells.

One approach to reduce polarity is to construct 3′,5′-cyclic phosphate analogs. The strategy is for the compounds to enter cells and be cleaved to 5′-monophosphate derivatives by phosphodiesterases. The free 5′-monophosphates can then be further phosphorylated to nucleoside di- and triphosphates. Some of these compounds, such as adenosine 3′,5′-cyclic phosphorothioates (4), rather than being catabolized to 5′-monophophate forms, have turned out to be inhibitors of nucleoside 3′,5′,-cyclic phosphodiesterases, or in other ways inhibit cell proliferation.

The charges on the 5′-phosphate moiety can also be reduced by alkyl blocking groups. Examples of active compounds include certain

0097–6156/89/0401–0124$06.00/0

nucleoside 5′-phosphonoformate (PFA) derivatives of 5-substituted 2′-deoxyuridines (5) and purines (6) that inhibit herpes simplex viruses. For each of the reported active compounds, the authors felt that cleavage of the PFA group occurred, and that the antiviral activity was due either to PFA or to a combination of PFA and the nucleoside analog, which in its free state would inhibit the virus anyway.

Phosphate mimics of nucleoside 5′-monophosphates such as 5′-sulfates and 5′-sulfamates (1) are non-polar and will readily penetrate into cells. The first compound to be identified in this series was nucleocidin, a naturally occurring antibiotic that possesses anti-parasitic activity (7). Nucleocidin is an inhibitor of protein synthesis, both of the host cell and the parasite (8), which makes it too toxic for human use. Adenosine 5′-sulfamate, a chemically-synthesized analog of nucleocidin, was shown to inhibit the growth of African trypanosomes (9) and to inhibit protein synthesis in *E. coli* (10). Adenosine 5′-sulfamate was devoid of antiviral activity when tested in our laboratory against the viruses listed in Table 1, and it had considerable cytotoxicity. The 5′-sulfamate analog of the antiviral agent tiazofurin was recently synthesized (11); it was also inactive in our viral screens. In a preliminary communication, certain uridine 5′-sulfamates were found to be selective inhibitors of African swine fever, and to a lesser extent, herpes simplex viruses *in vitro* (12).

The 5′-sulfamate functionality was also used to link nucleosides to hexose moieties in the preparation of a series of uridine diphosphate sugar analogs (13). The more active of these structures inhibited many enveloped viruses at 20 to 100 μg/ml (14). The antiviral activity was attributed to the inhibition of glycosylation of viral proteins and on DNA synthesis (14).

As part of the antiviral program at this institution, ribavirin 5′-sulfamate (5′-O-sulfamoyl-1-β-D-ribofuranosyl-1,2,4-triazole-3-carboxamide; see structure below) was synthesized using a scheme similar to that used to synthesize adenosine 5′-sulfamate (15). The antiviral activity and mode of action of this interesting new nucleotide analog are presented here.

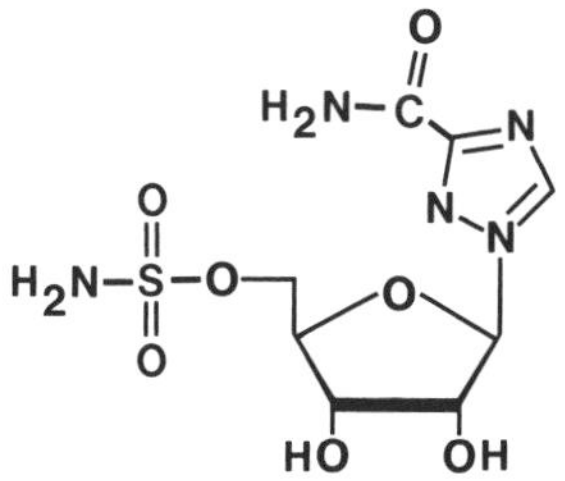

Antiviral activity in cell culture and in mice

Initially, ribavirin 5′-sulfamate was evaluated *in vitro* for antiviral activity in a primary screen (Table 1). The compound inhibited human cytomegalo-, herpes simplex, and parainfluenza virus-induced cytopathology (CPE) in a narrow concentration range, but had a more pronounced effect against Semliki Forest virus. At 10 μM, Semliki Forest virus-induced CPE was inhibited by 50%, whereas at 32 to 320 μM the suppression of CPE was 100%. This activity was unique since ribavirin was not active against Semliki Forest virus at $\leq$1000 μM. The analog showed no effect against influenza A and rhinoviruses, and it was toxic to the cells used to replicate these viruses at concentrations above 10 μM. In contrast, ribavirin markedly inhibited the replication of influenza A virus, and to a lesser extent inhibited rhinovirus 1-A and parainfluenza virus type 3. At antiviral concentrations, ribavirin 5′-sulfamate altered the morphology of Vero cells in stationary cell monolayers, indicating some partially toxic effects. In cell proliferation assays, actively replicating uninfected Vero cells were arrested in their growth by 50% at 1-3 μM.

Table 1. Inhibitory activities of ribavirin 5′-sulfamate and ribavirin against various viruses

Virus (Strain)	Cell Line	ED50* (μM) / Virus Rating** Rib.5′-Sulf.	Ribavirin
Human cytomegalo (AD-169)	MRC-5	1/0.4	>1000/0.0
Herpes simplex 2 (G)	Vero	10/0.4	>1000/0.0
Influenza A (Chile)	MDCK	>10/0.0	32/1.2
Parainfluenza 3 (C243)	Vero	10/0.4	100/0.8
Rhino 1-A (2060)	HeLa	>10/0.0	320/0.4
Semliki Forest(Original)	Vero	10/0.8	>1000/0.0

*Fifty percent virus-inhibitory concentration, determined from cytopathic effect inhibition assay (16).
**Virus ratings calculated by the method of Sidwell (17). Ratings of $\leq$0.5 indicate weak antiviral activity; 0.6-0.9, moderate activity; 1.0 and above, marked activity.

The rest of the investigations were performed using Semliki Forest virus in Vero cells. The virus is a positive-stranded RNA virus classified as a member of the togavirus family (18). Togaviruses are responsible for causing many encephalitic diseases in humans throughout the world. It was thought that by understanding the compound's activity in some detail, the information might be helpful to design more selective inhibitors of these viruses.

The next experiment determined the effects of ribavirin 5′-sulfamate on Semliki Forest virus replication by quantifying extracellular virus yield by plaque reduction assay methods (19). This was done under conditions where different multiplicities of infecting virus (MOI) were used (MOI refers to the virus to cell ratio). When the virus and compound were added simultaneously to cell monolayers (Table 2), a reduction of $\geq$1 log of virus production was achieved at 50-100 μM concentrations, regardless of the MOI. However, even at 100 μM the degree of inhibition of virus yield decreased with increasing MOI.

The virus destroyed the cell monolayer in 12 hours at an MOI of 10 and 48 hours at an MOI of 0.01. It usually takes a nucleoside or nucleotide analog about 2 to 6 hours to equilibrate into cells, and

Table 2. Antiviral activity of ribavirin 5′-sulfamate under varying multiplicities of Semliki Forest virus infection

Inhibitor Conc.(μM)	Log_{10} Plaque Forming Units/ml					
	Simultaneous Treatment**				Pre-Treatment* Actinomcyin D Present	
					None	(5 μM)
	0.01**	0.1	1	10	10	10
0	9.3	9.3	9.7	8.3	9.4	9.5
6.25	9.5	9.0	8.6	8.6	9.4	9.0
12.5	9.6	8.9	8.6	8.3	7.7	9.1
25	9.3	8.7	8.3	7.5	6.6	6.9
50	5.6	6.8	6.7	7.3	5.6	5.9
100	3.5	4.6	5.7	6.3	5.2	5.3
Log reduction at 100 μM	5.8	4.7	4.0	2.0	4.2	4.2

*Ribavirin 5′-sulfamate was present 18 hrs before and during virus replication. Actinomycin D was added with the infecting virus.
**Ribavirin 5′-sulfamate and virus were added simultaneously to the cells.
***Multiplicity of virus infection (MOI).
SOURCE: Reproduced with permission from ref. 23. Copyright 1988 Elsevier.

if necessary be metabolized (20,21). This means that at the high MOI the virus was well on its way to destroy the cell culture before complete equilibration of the inhibitor occurred. To compensate for this, the cells were pre-treated with ribavirin 5′-sulfamate (Table 2). The experiment was conducted in the presence or absence of actinomycin D, since actinomycin D is an important tool needed later in the mode of antiviral action studies. By starting treatment before virus inoculation, ribavirin 5′-sulfamate caused a more pronounced inhibition of virus yield. Actinomycin D did not cause a reversal (antagonism) of the antiviral activity of the analog at 25-100 μM. In contrast, actinomycin D is known to reverse ribavirin's activity against the viruses it inhibits (21,22).

In animal experiments ribavirin 5′-sulfamate was first tested in uninfected mice to determine its maximum tolerated dose. By dividing the daily dose, mice tolerated the analog at ≤40 mg/kg for 7 days. In an experimental infection (Table 3), most mice treated daily with 20 and 40 mg/kg survived an otherwise lethal encephalitic infection with Semliki Forest virus. Less of a protective effect was achieved at 10 mg/kg, although the increase in survivors was still statistically significant. The compound induced severe weight loss at 40 mg/kg and some loss at 20. With each day of treatment the uninfected control mice became more ill due to drug toxicity, as judged by weight loss and their overall appearance. Even the mice treated at 10 mg/kg appeared less healthy than saline-treated controls. A subsequent study showed that doses below 10 mg/kg provided no protection to infected mice.

Mode of antiviral action

The effects of ribavirin 5′-sulfamate on the incorporation of [3-H]-uridine and [3-H]amino acids into acid-precipitable macromolecules

Table 3. Effects of ribavirin 5′-sulfamate on a lethal Semliki Forest virus infection in mice

Dose* (mg/kg)	Survivors/ Total (%)	Toxicity Controls % Weight Difference** Day 7	Day 14
0	0/12 (0)	0	0
10	7/12 (58)***	-1	+2
20	11/12 (92)***	-13	-6
40	10/12 (83)***	-32	-13

*Compound or saline was administered twice a day in a divided dose for 7 days starting 2 hours before virus inoculation.

**The percent difference in mouse weight between the untreated and treated uninfected mice.

***Statistically significant ($p<0.01$), determined by the two-tailed Fisher exact test.

SOURCE: Reproduced with permission from ref. 23. Copyright 1988 Elsevier.

of uninfected and Semliki Forest virus-infected Vero cells was determined. These standard assays (19) are an indication of effects on RNA and protein syntheses, respectively. The intracellular, unincorporated (acid-soluble) amounts of these and all subsequently used radioactive compounds were also determined (19). Unless otherwise mentioned, treatment of cells with ribavirin 5′-sulfamate did not alter the uptake of these molecules into acid-soluble pools relative to untreated cells. In uninfected cells, a dose-dependent inhibition of amino acid incorporation occurred, with >50% inhibition at 6 µM (Table 4). The effect of ribavirin 5′-sulfamate on uridine incorporation was not dose-dependent (a 30% inhibition was observed at all concentrations tested). An overall 30% inhibition in acid-soluble counts also occurred at the same concentrations, which indicate a moderate inhibition of [3-H]uridine uptake into the cell. These results show that ribavirin 5′-sulfamate inhibited cellular protein synthesis, but had only a weak effect on RNA synthesis.

In virus-infected cells, the activity of the analog was studied in the presence or absence of actinomycin D (Table 4). Actinomycin D was used to suppress cellular mRNA synthesis and subsequent protein expression without interfering with the corresponding viral processes. Ribavirin 5′-sulfamate inhibited amino acid incorporation in infected cells to about the same degree whether actinomycin D was present or not. This was slightly less than the degree of inhibition observed in uninfected cells. For this reason it was not possible to conclude at this stage of experimentation that the analog inhibited viral protein synthesis.

A differential effect was noted between the incorporation of uridine into infected cells treated with actinomycin D compared to cells receiving no actinomycin D (Table 4). In the absence of actinomycin D, an overall 50% inhibition of incorporated uridine occurred at 6 to 100 µM, which was similar in nature to the lack of dose-responsiveness of ribavirin 5′-sulfamate in uninfected cells. However, in infected cells treated with actinomycin D where primarily viral RNA synthesis took place, a marked suppression of uridine incorporation was evident at >12.5 µM. These data indicate that ribavirin 5′-sulfamate interfered with viral RNA synthesis.

Table 4. Effects of ribavirin 5′-sulfamate on incorporation of uridine and amino acids into Semliki Forest virus-infected and uninfected cells

	Percent of Untreated Control					
	Uninfected		Virus-Infected			
Inhibitor* Conc.(μM)	Uridine	Amino Acids	Uridine	Uridine +Act.D**	Amino Acids	Amino Acids +Act.D**
6.25	72	37	49	54	65	80
12.5	69	26	42	16	47	39
25	71	21	44	9	38	37
50	69	18	46	10	27	23
100	68	11	54	9	16	9

*Treatment with ribavirin 5′-sulfamate before and during radio-labeling (with [3-H]uridine or [3-H]amino acids) was as follows: uninfected cells, 22 hrs pre-treated/2 hrs labeled; infected cells, 18 hrs pre-treated/4 hrs incubation of virus and inhibitor/2 hrs labeled.

**Actinomycin D (5 μM) was added at the time of virus infection.

SOURCE: Reproduced with permission from ref. 23. Copyright 1988 Elsevier.

In order to determine whether viral protein synthesis was inhibited in treated cultures, viral and cellular polypeptides were separated and visualized by PAGE/fluorography methods (24,25). Infected cells devoid of inhibitor exhibited distinct new bands of protein (lane 2 of Figure 1) not present in uninfected cultures (lane 1). In lanes 3-5, treatment with ribavirin 5′-sulfamate at 25-100 μM completely blocked the expression of these viral proteins. Also, the intensity of cellular proteins in these lanes was somewhat diminished relative to the untreated, uninfected control. These results show that ribavirin 5′-sulfamate inhibited viral and cellular protein synthesis, but not necessarily to the same degree under these treatment conditions.

Since the above studies demonstrated that ribavirin 5′-sulfamate inhibited viral and cellular protein synthesis and viral RNA synthesis, it was hypothesized that the inhibition of viral RNA synthesis was a consequence of the inhibited expression of the viral RNA polymerase protein. An alternate hypothesis is that ribavirin 5′-sulfamate directly inhibited the function of the viral RNA polymerase. To discriminate between the two hypotheses, viral RNA polymerase activity was partially purified from cells (26). The viral polymerase located in the cellular P15 fraction (mitochondria and membranes) was quantified from treated and untreated infected cells (Table 5). In these enzyme assays, the RNA polymerase reaction mixtures (26) did not contain ribavirin 5′-sulfamate. The results show that the amount of enzymatic activity recovered from inhibitor-treated cells was a function of the concentration present during the time the virus was replicating in cell culture. When 1 mM ribavirin 5′-sulfamate was added directly to a viral RNA polymerase reaction obtained from untreated, virus-infected cells, only a 27% decrease in the rate of polymerization occurred, indicating that the analog was only a weak inhibitor of the enzyme. These results demonstrate that viral RNA synthesis was inhibited

Table 5. Effects of ribavirin 5′-sulfamate on the amount of Semliki Forest virus RNA polymerase activity recovered from infected cells

Inhibitor* Conc.(μM)	Polymerase Activity**	Percent of Control
0	3,698	100
6.25	1,673	45
12.5	670	18
25	148	4
50	0	0
100	0	0

*Inhibitor was present 18 hours before and during the virus replication period (4 hrs) in cell culture, but was absent from the RNA polymerase reaction.
**Expressed as counts per minute per million cells.
SOURCE: Reproduced with permission from ref. 23. Copyright 1988 Elsevier.

as a consequence of the inhibition of viral protein expression (i.e., the viral RNA polymerase was absent from treated cells due to the inhibited translation of its mRNA).

Inhibition of protein translation and DNA synthesis

A series of studies was conducted to find the specific intracellular site of action of ribavirin 5′-sulfamate. Using a commercially available rabbit reticulocyte lysate system capable of translating mRNA to protein, test tube reactions were conducted to evaluate the effect of the analog on protein translation in vitro (Table 6). Puromycin was also evaluated simultaneously as a positive control (27). Ribavirin 5′-sulfamate inhibited protein translation in this system in a dose-dependent manner from 15-100 μM. Puromycin was also active, and proved to be a more potent inhibitor of the reaction. When the two inhibitors were present in the same reaction mixture (Table 6), an additive drug effect occurred. This could result from an interaction of the compounds at the same or different sites on the ribosomal complex.

Translation of mRNA to protein can be inhibited at many steps. For example, puromycin interacts at the aminoacyl acceptor site on ribosomes, and is incorporated into the polypeptide chain where it terminates it (27). The result is the formation of incomplete polypeptide fragments. To see whether ribavirin 5′-sulfamate would have a similar effect, polypeptides formed in the presence of each inhibitor were visualized using electrophoresis/fluorography methods (24,25). The results of the assay show that puromycin caused only small molecular weight polypeptide fragments to be produced (lane 5 of Figure 2) with a maximum molecular weight of 40,000. On the other hand all sizes of polypeptides were synthesized in ribavirin 5′-sulfamate treated cells (lane 3) but the rate of synthesis was diminished relative to the untreated control (lane 2), based upon the intensity of the polypeptide bands in the gel. These results indicate that ribavirin 5′-sulfamate and puromycin do not act in the same way to inhibit protein translation, since ribavirin 5′-sulfamate did not chain-terminate polypeptide synthesis.

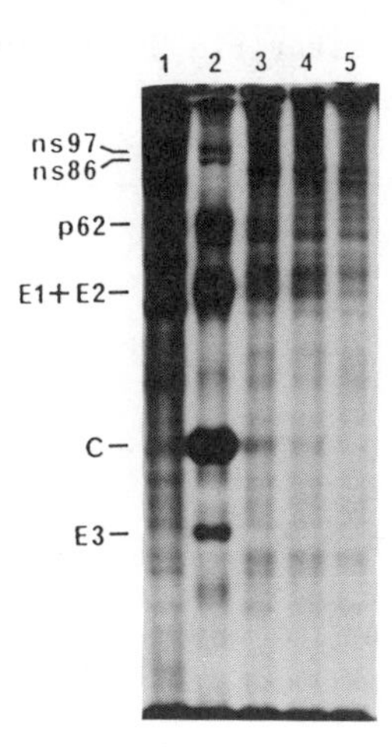

Figure 1. Effects of ribavirin 5'-sulfamate on viral and cellular polypeptide synthesis. Treatment with the inhibitor began 18 h before virus; cells were exposed to [3-H]amino acids 4–6 h after virus adsorption. Lane 1, uninfected untreated; lanes 2–5, infected and treated with 0, 25, 50, and 100 μM inhibitor, respectively. Designations at left indicate positions in lane 2 where viral-specific nonstructural (ns86 and ns97) and structural (C, capsid; E1, E2, and E3, glycoproteins; p62, precursor to E1–E3) polypeptides occur. (Reproduced with permission from ref. 23. Copyright 1988 Elsevier.)

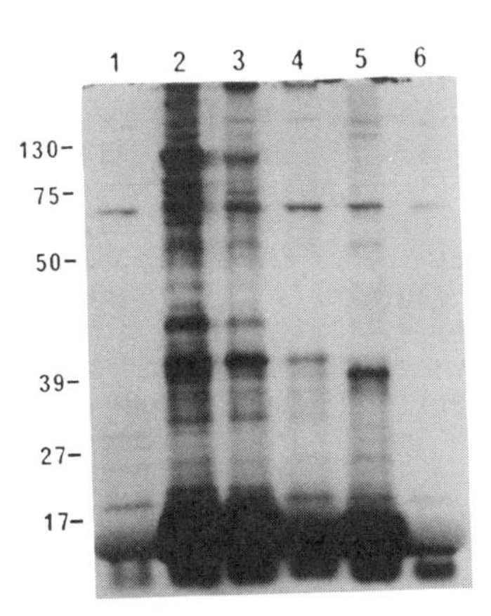

Figure 2. Inhibition of proteins translated by a rabbit reticulocyte lysate system. Lane 1, background control (mRNA absent); lane 2, untreated control; lanes 3–4, ribavirin 5'-sulfamate at 30 and 300 μM, respectively; lanes 5–6, puromycin at 3 and 30 μM, respectively. Molecular weight markers appear at left of figure. (Reproduced with permission from ref. 23. Copyright 1988 Elsevier.)

Table 6. Effects of ribavirin 5′-sulfamate and puromycin on protein translation* *in vitro* (23)

Inhibitor	Conc.(μM)	Percent of Control
Ribavirin 5′-Sulfamate	10	102
	15	76
	30	45
	100	14
Puromycin	1	76
	3	21
	10	0
Rib.5′-Sulf./Puromycin	15/1	52 (58)**
	30/1	37 (34)

*Reactions contained a rabbit reticulocyte lysate system, rabbit globin mRNA, and [35-S]methionine.
**( ) = expected value for additive drug interaction.

Once it was known that ribavirin 5′-sulfamate inhibited protein translation, it became evident that the compound may be similar to adenosine 5′-sulfamate in its action. Initially it was thought that the two compounds had dissimilar modes of action, because adenosine 5′-sulfamate was inactive against Semliki Forest virus in cytopathic effect inhibition assays. From the literature, adenosine 5′-sulfamate was shown to inhibit aminoacyl-tRNA synthetase reactions (10), which are very early steps required in protein synthesis. Aminoacyl-tRNA synthetases link amino acids to respective tRNA's, which then carry the amino acids to the ribosome complex. The experiment in Table 7 shows that ribavirin 5′-sulfamate inhibited methionine-tRNA synthetase (the only synthetase that was tested; there is at least one aminoacyl-tRNA synthetase for each amino acid).

The experimental method (10) used to derive the data in Table 7 may not be obvious to the reader so will be explained. The protein translation experiment in Table 6 reports total inhibition of polypeptide synthesis and of aminoacyl-tRNA synthesis. This is because radioactive proteins and radioactive aminoacyl-tRNA's are all trapped onto filters counted in the assays. Referring to Table 7, by boiling the filters in hot perchloric acid (PCA), the aminoacyl-tRNA's degrade and elute off. Thus, the difference in counts from boiled versus unboiled filters represents activity associated with aminoacyl-tRNA synthesis. Ribavivin 5′-sulfamate inhibited total synthesis, polypeptide synthesis, and aminoacyl-tRNA synthesis all to approximately the same degree. Adenosine 5′-sulfamate was also inhibitory to this process, and by further experimentation was shown not to inhibit the elongation of polypeptides in the presence of pre-formed aminoacyl-tRNA's (10). By inference the same may be true for ribavirin 5′-sulfamate.

DNA synthesis is a process that is dependent upon protein synthesis (28), so one would expect ribavirin 5′-sulfamate to inhibit that process. This was confirmed by the experimental results of Table 8, using [3-H]thymidine incorporation as an indicator. Some of the antiproliferative and anti-DNA virus activities of the analog are related to inhibition of DNA synthesis, but this probably would not interfere with Semliki Forest virus replication.

Table 7. Effects of ribavirin 5′-sulfamate on protein translation of rabbit globin mRNA and on aminoacyl-tRNA synthesis

	Percent of Untreated Control					
	10 Minute Reaction*			60 Minute Reaction*		
Inhibitor Conc.(μM)	Cold PCA	Hot PCA**	Cold Minus Hot PCA	Cold PCA	Hot PCA	Cold Minus Hot PCA
30	40	50	37	30	36	28
100	13	10	14	14	17	13
300	8	1	9	5	7	5

*A rabbit reticulocyte lysate system was run using [35-S]methionine as indicator for the reactions.
**Filters were boiled in 3.5% perchloric acid (PCA) to degrade the aminoacyl-tRNA′s.

Inhibition of methylation, sulfuration, and polyamine synthesis

The structure of ribavirin 5′-sulfamate potentially mimics certain naturally occurring 5′-thioadenosine derivatives such as S-adenosylmethionine (AdoMet), S-adenosylhomocysteine (AdoHcy), methylthioadenosine, and 3′-phosphoadenosine-5′-phosphosulfate (PAPS) that play important roles in cellular metabolism. The next set of experiments show the effects of ribavirin 5′-sulfamate to inhibit processes that these molecules are involved in.

AdoMet serves as a methyl donor in cellular reactions. The inhibition of AdoHcy hydrolase leads to an accumulation of AdoHcy, which inhibits methylation reactions (29). Specific inhibitors of AdoHcy hydrolase have been shown to possess antiviral activity (30), because certain viruses components require methylation. In Table 8, cells were treated with ribavirin-5′-sulfamate and labeled with [3-H-(methyl)]AdoMet. A marked dose-dependent inhibition of methyltransfer from AdoMet to macromolecules occurred. This effect could have been caused by inhibition of enzymes in the AdoMet pathway such as AdoHcy hydrolase, or as a general consequence of protein synthesis inhibition, or both.

Table 8. Effects of ribavirin 5′-sulfamate on various cellular functions

	Percent of Untreated Control			
Inhibitor* Conc.(μM)	Thymidine Incorp.	Sulfate Incorp.	Methyl Transfer	Methionine Incorp.
6.25	48	35	47	33
12.5	22	16	31	21
25	15	10	24	15
50	10	6	19	12
100	8	3	13	9

*Cells were treated with ribavirin 5′-sulfamate before and during labeling as follows: [3-H]thymidine, 22 hrs pre-treated/2 hrs labeled; [35-S]sulfate, 20 hrs pre-treated/4 hrs labeled; [3-H-(methyl)]AdoMet and [35-S]methionine, 5 hrs pre-treated/1 hr labeled.

Cells treated with 6-100 μM ribavirin 5′-sulfamate and labeled with [35-S]methionine showed approximately a 180% elevation in acid-soluble radioactivity at each concentration compared to untreated cells, which could represent an increase in AdoHcy. To investigate this possibility, extracts from these cells were analyzed by high pressure liquid chromatography (31). The results showed elevated amounts of methionine and AdoMet, and decreased amounts of cysteine and AdoHyc present in inhibitor-treated cells (Table 9), suggesting that ribavirin 5′-sulfamate did not inhibit AdoHcy hydrolase, but more generally affected methyltransfer reactions as a consequence of inhibiting protein synthesis. Also, protein synthesis (measured by [35-S]methionine incorporation, Table 8) was inhibited to nearly the same degree as the inhibition of methyltransfer reactions, indicating a link between the two processes. The buildup of methionine and AdoMet were probably the result of their lack of utilization for protein synthesis and methyltransfer reactions, respectively. The role that inhibition of methyltransfer reactions play in Semliki Forest virus replication is not known, although the related Sindbis virus is only weakly affected by inhibitors of AdoHcy hydrolase (32).

Table 9. Metabolites of [35-S]methionine in cells treated with ribavirin 5′-sulfamate

Compound	Counts per Minute per 10^5 Cells	
	Untreated	Rib.5′-Sulf.*
Methionine	13,438	23,785 (177)**
Cysteine	24,472	16,151 (66)
AdoHcy	1,417	864 (64)
AdoMet	31,039	56,180 (181)

*Inhibitor (6.25 μM) was present 5 hrs before and 1 hr during radiolabeling.
**( ) = percent of untreated control.

AdoMet also serves in the first step leading to the synthesis of polyamines, which play a role in cell proliferation (33). Cells treated with ribavirin 5′-sulfamate and cycloheximide (another inhibitor of protein synthesis, 28) were inhibited in spermidine and spermine syntheses (Table 10). This was determined by following the metabolism of [3-H]ornithine (34). Puromycin was not useful to show an inhibitory effect because at active concentrations that compound was overtly toxic. To try and discriminate whether the inhibition of polyamine synthesis was due to a direct activity of ribavirin 5′-sulfamate on the polyamine pathway or due to general protein synthesis inhibition, a parallel study was run showing inhibited incorporation of [3-H]methionine into proteins (Table 10). At 10 μM, ribavirin 5′-sulfamate was more inhibitory to protein synthesis than it was to polyamine synthesis, whereas at 30-100 μM both processes were equally inhibited. The same was generally true for cycloheximide at 30-100 μM. The data indicate that these inhibitors of protein synthesis indirectly affected polyamine synthesis as a consequence of inhibiting protein synthesis. The

Table 10. Effects of ribavirin 5′-sulfamate and cycloheximide on methionine incorporation and polyamine synthesis

Inhibitor*	Conc. (μM)	Percent of Untreated Control: Methionine Incorp.	Ornithine Metabolism to: Putrescine	Spermidine	Spermine
Ribavirin 5′-Sulfamate	10	24	107	61	33
	30	14	89	12	9
	100	8	95	6	5
Cycloheximide	10	85	91	253	141
	30	56	110	42	23
	100	17	107	11	4

*Cells were treated with inhibitor before and during labeling as follows: [3-H]methionine, 5 hrs pre-treated/1 hr labeled; [3-H]-ornithine, 6 hrs pre-treated/12 hrs labeled.

inhibition of polyamine synthesis would adversely affect the replication of DNA viruses, but probably would not lead to an inhibition of RNA viruses.

PAPS is a substrate for sulfuration of certain intracellular proteins containing tyrosine molecules (35) and for other structural components such as chondroitin sulfate. It was of interest to know the effects of ribavirin 5′-sulfamate on sulfuration reactions, and whether this could have any virus-inhibitory consequences. Other investigators have reported the activities of specific inhibitors of protein sulfurases, such as sodium chlorate (35), as well as compounds that inhibit sulfuration reactions non-selectively as a consequence of inhibiting protein synthesis (36). In the present studies, cells treated with ribavirin 5′-sulfamate had a decreased capacity to incorporate inorganic [35-S]sulfate into macromolecules (Table 8). By following this effect over time, the intracellular rate of sulfuration was enhanced during the first hour of treatment, then decreased in parallel with the inhibition of [3-H]amino acid incorporation into protein (Table 11). These results suggest that ribavirin 5′-sulfamate was acting non-specifically to inhibit the sulfuration reactions. The role that protein sulfuration may play

Table 11. Ribavirin 5′-sulfamate inhibition of sulfate and amino acid incorporations over time

Time (hrs)	Percent of Untreated Control*: Sulfate Incorp.	Amino Acids Incorp.
1	160	94
2	65	74
3	34	52
4	30	27
5	25	27
6	25	23

*Cells were pulse-labeled hourly using [35-S]sulfate or [3-H]amino acids. The inhibitor (25 μM) was added at 0 time.

on the replication of Semliki Forest virus was tested by treating cells with 10 mM sodium chlorate, where it was found to be inactive.

Uptake of ribavirin 5′-sulfamate into cells

The last questions to be addressed deal with the fate of the analog inside of mammalian cells. To what extent is ribavirin 5′-sulfamate taken up by cells, and what metabolites, if any are formed intracellularly? To answer these questions, extracts from treated cells were analyzed for the presence of the compound and possible metabolites by a reverse phase high pressure liquid chromatographic (HPLC) method. Taking into account the inherent variability of the assay, approximately the same concentrations of the inhibitor were present inside and outside the cells (Table 12), and no metabolites of the compound were detected. The sensitivity of the method, which relied upon an ultraviolet (UV) light detection system, did not allow the detection of minute quantities of potential metabolites, however (1 nanomole of compound was the minimum amount detectable by UV). The data from Table 11 is a bioassay of the time it took for the analog to equilibrate fully into cells to exert its maximal effect, which was about 4 hours. Presumably if active metabolites were being formed, the process to completely inhibit protein synthesis would have taken longer. For example, 12 hours were required to fully generate the antiviral metabolite ribavirin 5′-triphosphate in cells (21). These studies support the hypothesis that ribavirin 5′-sulfamate is solely responsible for antiviral and anticellular effects.

Table 12. Uptake of ribavirin 5′-sulfamate into cells*

Extracellular Conc. (μM)	Intracellular Conc. (μM) at	
	6 Hours	18 hours
100	137	71
300	368	233
1000	965	1005

*Cell extracts were assayed using a C-18 HPLC column. The elution time of ribavirin 5′-sulfamate (detected by UV at 210 nm) was 16 minutes at 1 ml/minute flow rate. Elution buffer was 0.02 M potassium phosphate, pH 3.5.

Conclusions and discussion

Ribavirin 5′-sulfamate inhibited Semliki Forest virus in cell culture and in mice, although it was considered to be partially toxic at concentrations or doses that provided protection. The mode of antiviral action of this nucleotide analog involved the inhibition of viral protein synthesis, which impacted on viral mRNA synthesis. All of the anticellular effects on DNA and polyamine syntheses, and on methyltransferase and sulfuration reactions were attributed to inhibition of cellular protein synthesis. The analog inhibited aminoacyl-tRNA synthesis, which was considered to be the major target for its action. Since no metabolites of the parent compound were detected in cells (within the limits of detectability by UV spectroscopy), ribavirin 5′-sulfamate itself appears to be responsible for all of the effects reported here. The compound fits into a larger class of protein translation inhibitors, including puromycin and cycloheximide (18), that will inhibit virus replication. Semliki Forest virus (and probably other positive stranded viruses), which requires an initial translation of its genome before the RNA can amplify, is particularly sensitive to these types of inhibitors.

In a recent report it was shown that aminoacyl-tRNA synthetases possess phosphotransferase activities (37). They interconvert AMP to ADP to ATP and vice versa. It is hypothesized that ribavirin 5′-sulfamate interacts at the AMP binding site to inhibit this class of enzymes. Its action is similar to that of adenosine 5′-sulfamate (10) which also inhibits aminoacyl-tRNA synthetases. Due to the much greater cytotoxicity of adenosine 5′-sulfamate, the antiviral activity of that compound was masked, leading to the conclusion that it was inactive as a virus inhibitor. In this regard ribavirin 5′-sulfamate was more selective, because cells were spared overt toxicity which allowed the antiviral effect to be visualized.

In the electrophoresis studies, viral polypeptide synthesis was completely inhibited whereas cell proteins were partially expressed. This differential effect probably relates to the fact that only a finite and small amount of virus message was present for translation, whereas the host mRNA was in greater abundance and was continually being produced. Competition for the ribosomes would favor host cell mRNA in this circumstance. It is only after the viral RNA polymerase and other key early proteins can be sufficiently expressed that the quantity of viral mRNA would increase inside infected cells.

Other kinds of viruses than those reported here may also be inhibited by the compound, especially positive-stranded viruses, but further investigation is probably unwarranted since the antiviral potential of this substance appears to be low. This was substantiated by the overt toxicity manifested in the animal study (Table 3). To follow up on the antiviral activity reported against human cytomegalovirus in vitro (Table 1), an experiment was conducted against murine cytomegalovirus in mice. In that test, ribavirin 5′-sulfamate was inactive. Otherwise, the nucleotide analog has not been evaluated in any other animal virus infection model.

Whether nucleoside 5′-sulfamates will find their way into the arsenal of compounds with clinical potential remains to be determined. Clearly, inhibitors need to be identified which have virus-specific modes of action, rather than affect cellular processes that viruses also depend on. Along these lines, it will be of interest to learn the mode of action of the recently reported uridine 5′-sulfamate inhibitors of African swine fever and herpes simplex viruses (12). The results of the present study show that inhibition of protein synthesis led to antiviral activity, which is not a unique observation. However, ribavirin 5′-sulfamate is the first compound reported to be virus-inhibitory as a consequence of inhibiting aminoacyl-tRNA synthesis.

Literature cited

1. Robins, R. K. Pharmaceut. Res. 1984, 1, 1-50.
2. Martin, J. C.; Verheyden, J. P. H. Nucleosides Nucleotides 1988, 7, 365-374.
3. Fuertes, M.; Witkowski, J. T.; Streeter, D. G.; Robins, R. K. J. Med. Chem. 1974, 17, 642-645.
4. Pereira, M. E.; Segaloff, D. L.; Ascoli, M.; Eckstein, F. J. Biol. Chem. 1987, 262, 6093-6100.
5. Griengl, H.; Hayden, W.; Penn, G.; De Clercq, E.; Rosenwirth, B. J. Med. Chem. 1988, 31, 1831-1839.
6. Vaghefi, M.; McKernan, P. A.; Robins, R. K. J. Med Chem. 1986, 29, 1389-1393.
7. Tobie, E. J. J. Parasitol. 1957, 43, 291-293.
8. Florini, J. R.; Bird, H. H.; Bell, P. H. J. Biol Chem. 1966, 241, 1091-1098.
9. Jaffee, J. J.; McCormack, J. J.; Meymerian, E. Exp. Parasitol. 1970, 28, 535-543.
10. Bloch, A.; Coutsogeorgopoulos, C. Biochem. 1971, 10, 4394-4398.
11. Andres, J. I.; Garcia-Lopez, M. T.; De las Heras, F. G.; Mendez-Castrillon, P. P. Nucleosides Nucleotides 1986, 5, 423-429.
12. Perez, S.; Fiandor, J.; Perez, C.; Garcia-Lopez, M. T.; Garcia-Gancedo, A.; De las Heras, F. G.; Fernandez, C. G.; Vilas, P.; Mendez-Castrillon, P. P. VII Internatl. Congr. of Virology, 1987, p 222.
13. Camarasa, M.-J.; Fernandez-Resa, P.; Garcia-Lopez, M. T. De las Heras, F. G.; Mendez-Castrillon, P. P.; Alarcon, B.; Carrasco, L. J. Med. Chem. 1985, 28, 40-46.
14. Gil-Fernandez, G.; Perez, S.; Vilas, P.; Perez, C.; De las Heras, F. G.; Garcia-Gancedo, A. Antiviral Res. 1987, 8, 299-310.
15. Shuman, D. A.; Robins, R. K.; Robins, M. J. J. Am. Chem. Soc. 1969, 91, 3391-3392.
16. Smee, D. F.; McKernan, P. A.; Nord, L. D.; Willis, R. C.; Petrie, C. R.; Riley, T. M.; Revankar, G. R.; Robins, R. K.; Smith, R. A. Antimicrob. Agents Chemother. 1987, 31, 1535-1541.
17. Sidwell, R. W. In Chemotherapy of Infectious Diseases; Gadebusch, H. H., Ed.; CRC: Cleveland, Ohio, 1976; pp 31-53.
18. Schlesinger, M. J.; Kaariainen, L. In The Togaviruses; Schlesinger, R. W., Ed.; Academic: New York, 1980; pp 371-392.
19. Smee, D. F.; Martin, J. C.; Verheyden, J. P. H.; Matthews, T. R. Antimicrob. Agents Chemother. 1983, 23, 676-682.
20. Votruba, I.; Holy, A.; De Clercq, E. Acta Virol. 1983, 27, 273-276.
21. Smee, D. F.; Matthews, T. R. Antimicrob. Agents Chemother. 1986, 30, 117-121.
22. Malinsoki, F.; Stollar, V. Virology 1980, 102, 473-476.
23. Smee, D. F.; Alaghamandan, H. A.; Kini, G. D.; Robins, R. K. Antiviral Res. 1988, 7 (in press)
24. Laemmli, U. K. Nature (London) 1970, 227, 680-685.
25. Bonner, W. M.; Lasky, R. A. Eur. J. Biochem. 1974, 46, 83-88.
26. Ranki, M.; Kaariainen, L. Virology 1979, 98, 298-307.
27. Nathans, D. Proc. Natl. Acad. Sci. USA 1964, 51, 585-592.

28. Seki, S.; Mueller, G. C. Biochim. Biophys. Acta 1975, 378, 354-362.
29. Glazer, R. I.; Hartman, K. D.; Knode, M. C.; Richard, M. M. Chiang, P. K.; Tseng, C. K. H.; Marquez, V. E. Biochem. Biophys. Res. Commun. 1986, 135, 688-694.
30. De Clercq, E. Biochem. Pharmacol. 1987, 36, 2567-2575.
31. Wagner, J.; Danzin, C.; Mamont, P. J. Chromatogr. 1982, 227, 349-368.
32. De Clercq, E.; Montgomery, J. A. Antiviral Res. 1983, 3, 17-24.
33. Pegg, A. E.; Coward, J. K.; Talekar, R. R.; Secrist, J. A. Biochem. 1986, 25, 4091-4097.
34. Seiler, N.; Knodgen, B. J. Chromatogr. 1980, 221, 227-235.
35. Baeuerle, P. A.; Huttner, W. B. Biochem. Biophys. Res. Commum. 1986, 141, 870-877.
36. Huttner, W. B.; Lee, R. W. H. J. Cell Biol. 1982, 95, 389a.
37. Rapaport, E.; Pemy, P.; Kleinkauf, H.; Vater, J.; Zamecnic, P. C. Proc. Natl. Acad. Sci. USA 1987 84, 7891-7895.

RECEIVED January 4, 1989

Chapter 10

Design, Synthesis, and Antiviral Activity of Nucleoside and Nucleotide Analogues

Victor E. Marquez

Laboratory of Medicinal Chemistry, National Cancer Institute, National Institutes of Health, Bethesda, MD 20892

Several categories of antiviral compounds have been designed by exploiting important differences in viral and cellular biochemistry, as well as differences in chemical stability and reactivity between parent "lead" molecules and their modified structures. The synthesis, antiviral activity and mechanism of action of two carbocyclic nucleoside analogues related to the natural product neplanocin A are discussed. While the cytosine carbocycle, cyclopentenyl cytosine (CPE-C, **3**), requires activation to form the biologically active nucleotide 5′-triphosphate analogue, a more selective antiviral effect is realized in the purine series by the modified 3-deazaneplanocin A (**6c**) analogue, due to its resistance to form nucleotide metabolites. In the area of dideoxy nucleoside analogues with anti-HIV activity, dideoxy-fluoro-ara-adenosine (ddF-ara-A, **15**) and dideoxy-fluoro-ara-inosine (ddF-ara-I, **16**), were designed as hydrolytically stable compounds. Although the beta stereochemistry of the fluorine atom proved not to be essential for chemical stability, it was critical for retaining the potent anti-HIV activity of the parent compounds. In the area of nucleotide analogues which incorporate in their structure an isosteric phosphate group, the complete synthesis of the phosphonate analogue of adenosine-2′-phosphate (**26**) is discussed. This compound was designed as a hydrolytically stable component of a 2′,5′-oligoadenylate trimer analogue (**17**) whose synthesis is in progress.

Nucleotide analogues can be divided into three categories: 1) base-modified, 2) sugar-modified, and 3) phosphate-modified. Most of the commonly known antiviral agents belong to the first two groups, which are generated intracellularly from their corresponding nucleoside precursors by either cellular or virus-coded enzymes. For the most part, these nucleotides exert their antiviral activity as the corresponding mono-, or triphosphate forms (e.g., ribavirin (1) and acyclovir (2)) which interfere with critical metabolic steps required for viral infectivity (See Figure 1). However, since these phosphorylated forms are also responsible for the toxicity observed against normal cells, their selective generation in just virally infected cells, or their specific interaction with key viral processes, become important factors in achieving a good therapeutic ratio.

The third category comprises those nucleotide analogues in which a modified phosphate group, or its equivalent, is already incorporated into the molecule (e.g, DHPG phosphonate) (3). These compounds are intended to overcome resistance from viruses which either naturally or by mutation do not have the capacity to phosphorylate the nucleoside substrates. Increase in their cellular uptake, which becomes a limiting factor, is achieved by the partial or total masking of the associated charge.

At the other end of the spectrum there are some antiviral nucleosides for which the desired antiviral effect can only be uncovered when formation of the nucleotide metabolites is prevented. In these cases, the antiviral selectivity is related to the specific interaction of the nucleoside, or a non-phosphorylated metabolite of it, with a vital viral process (e.g. neplanocin A) (4).

The design, synthesis, and antiviral activity of several new antiviral agents that fall into these different classes will be discussed.

Cyclopentenyl Cytosine (3, CPE-C).

Chemistry. Cyclopentenyl cytosine (**3**, CPE-C, Scheme 1) is a pyrimidine analogue of the fermentation product, neplanocin A. The compound was independently synthesized in our laboratory (5,6) and that of Ohno in Japan (7,8) by two distinct chemical approaches that centered around the stereospecific generation of the chiral cyclopentenyl amine precursor **1**. A simplified version of our most recent approach to CPE-C is shown in Scheme 1 (6). Limited structure-activity studies have indicated that the integrity of the carbocyclic component of CPE-C is essential for antiviral activity. For example, the corresponding 2′-deoxy and ara-CPE-C analogues were ineffective as antiviral agents with only the ara isomer showing some activity against influenza (9). Interestingly, both analogues were also devoid of the excellent antitumor properties characteristic of CPE-C (10).

Antiviral Activity. CPE-C has demonstrated significant antiviral activity against several DNA viruses with high virus ratings (VR = 2.7 - 4.6) at doses ranging from 0.1 to 2.7 μg/ml (6). Of particular interest was the effect that this drug displayed against the thymidine kinase deficient strain of HSV-1 (6,11). The antiviral activity of CPE-C was also evidenced *in vivo* as demonstrated by the results in the murine vaccinia tailpox model where it significantly reduced the development of virus-induced tail lesions (Table I) (12). In this assay, CPE-C was evaluated

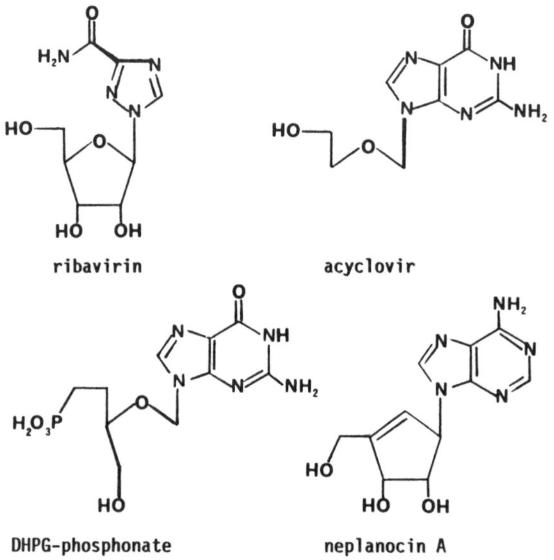

Figure 1. Chemical structures of selected antiviral agents

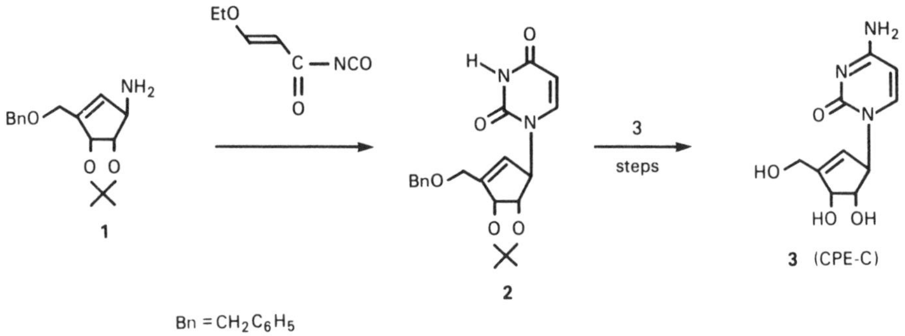

Scheme 1. Synthesis of CPE-C.

for efficacy against the IHD strain of vaccinia in mice, challenged with the virus by the intravenous route (tail vein).

Table I. Antiviral Activity of CPE-C
(Murine Vaccinia Tailpox Model)

PBS Control[a]	Ara-A		CPE-C	
Pox Count	Dose[b] (mg/kg)	Pox Count[c]	Dose[b] (mg/kg)	Pox Count[c]
45.8	300	2.8	1.5	0.8

[a] Diluent-treated mice (phosphate buffer saline) challenged with the virus
[b] QD 1-7 started on the day preceding virus challenge
[c] Mean value (20 mice)

Other DNA viruses sensitive to CPE-C were cytomegalovirus and varicella-zoster, as demonstrated in the yield reduction assay and plaque reduction assay, respectively (6).

CPE-C also demonstrated significant activity against a spectrum of RNA viruses *in vitro* (6). Virus ratings greater than two were obtained against vesicular stomatitis, Punta Toro and Hong Kong viruses. However, the utility of these activities is moderated by low therapeutic index values. By contrast, this does not appear to be the case for the Japanese encephalitis virus, where CPE-C showed a VR of 2.4, with greater potency and higher therapeutic index than ribavirin (6). No antiretroviral activity against HIV was observed for CPE-C in ATH8 cells.

Mechanism of Action. CPE-C is converted to the nucleotide triphosphate analogue of cytidine (CPE-CTP), and as such it behaves as a powerful inhibitor of CTP synthetase in cells (13). Although a direct correlation between the inhibition of this enzyme and its antiviral activity has not been demonstrated, it appears, as suggested by De Clercq et al., that this rate-limiting enzyme in the *de novo* pyrimidine pathway may be a therapeutically exploitable target for some viruses (11). The remarkably broad spectrum of activity observed for this compound indirectly supports this assumption.

3-Deazaneplanocin A (**6c**).

Chemistry. This compound was prepared by a similar approach to the simplified version of the synthesis of neplanocin A that we had earlier developed (14). Thus, the cyclopentenyl mesylate **4** was reacted with the sodium salt of 4-chloro-imidazo[4,5-*c*]-pyridine (6-chloro-3-deazapurine) to give a mixture of both N-1 and N-3 isomers (Scheme 2). After chromatographic separation and removal of the protective groups in each isomer, the major component was identified as the desired compound (**6a**) by NMR spectroscopy and diagnostic NOE measurements. The chloropurine intermediate **6a** was then converted in two steps to 3-deazaneplanocin A which was isolated as a crystalline solid. The molecular structure of 3-deazaneplanocin A was also confirmed by X-ray crystallography (15).

Antiviral Activity. 3-Deazaneplanocin A was first evaluated *in vitro* against viruses for which neplanocin A and other adenosine analogues are known to be particularly effective (16,17). These included a DNA virus (vaccinia) and two (-)RNA viruses (parainfluenza and vesicular stomatitis). In Table II, the minimum inhibitory concentration capable of reducing virus-induced cytopathogenic effects by 50% (ID_{50}), virus rating (VR), and therapeutic index (TI), are compared for 3-deazaneplanocin A and three positive control standards for each of the viruses. Treatment with 3-deazaneplanocin A resulted in a marked inhibition of parainfluenza type 3 virus (Huebner C243 in H.Ep-2 cells) and Vesicular stomatitis (VSV Indiana in L929 cells). The activity against VSV is particularly striking when compared with the structurally related standard 3-deazacarbocyclic adenosine analogue (3-deaza-C-Ado) (18) with which 3-deazaneplanocin is structurally related. With other RNA viruses, such as RSV (respiratory syncytial) and human influenza (type Ao/PR/8/34), however, 3-deazaneplanocin A did not show any promising antiviral selectivity. Similarly, against a number of picornaviruses, the compound was marginally effective. The compound resulted inactive against rhinovirus and Coxsackie type B1, and only a marginal effect was observed on Coxsackie type A21. The inhibition of poliovirus-induced cytopathogenic effects was also variable and not dose-responsive. Against a DNA virus, such as vaccinia, 3-deazaneplanocin A demonstrated a remarkable effect. These results prompted the *in vivo* study of the drug against vaccinia which is summarized in Table III. Treatment with 3-deazaneplanocin A, administered subcutaneously once daily from day -1 through day +5, resulted in a significant reduction in vaccinia induced tailpox compared to diluent-treated control mice. The therapeutic effect was evident at dose levels of 8 and 4 mg/Kg/day, although the higher dose resulted in weight loss in the toxicity control animals.

Mechanism of Action. Various adenine nucleoside analogues which function as inhibitors of *S*-adenosylhomocysteine hydrolase (AdoHcy-ase) have been identified as antiviral agents (19,20). Furthermore, a recent study has revealed the existence of a close correlation between their inhibitory potency against AdoHcy-ase and their antiviral activity (VSV *in vitro*) (17). In vaccinia, as well as in other sensitive viruses that require a methylated 5'-cap on their mRNAs, inhibition of AdoHcy-ase results in the marked elevation of AdoHcy levels and the consequent feedback inhibition of *S*-adenosylmethionine (AdoMet)-dependent methylation reactions. This enzyme is a unique target for the development of antiviral nucleoside analogues that does not require anabolism of the drugs

Table II. Antiviral Activity of 3-Deazaneplanocin A (In Vitro)

3-Deazaneplanocin A				Positive Control			
RNA Viruses	ID_{50} (μg/ml)	VR[a]	TI[b]	Compound	ID_{50} (μg/ml)	VR	TI
Parainfluenza type 3	0.05	3.6	2.0	Ribavirin	17.50	2.5	5.7
Vesicular Stomatitis	0.07	3.6	4.6	3-Deaza-C-Ado	2.6	3.2	3.8
DNA Viruses							
Vaccinia	0.32	3.1	3.1	Ara-A	2.0	3.7	16.3

[a] Virus rating (Shannon and Arnett, SoRI)
[b] Therapeutic index: minimum toxic concentration /ID_{50}

Table III. Antiviral Activity of 3-Deazaneplanocin-A (Murine Vaccinia Tailpox Model)

PBS Control[a]	Ara-A		3-Deazaneplanocin-A	
Pox Count	Dose[b] (mg/kg)	Pox Count[c]	Dose[b] (mg/kg)	Pox Count[c]
72.7	300	2.1	4	4.3

[a] Diluent-treated mice (phosphate buffer saline) challenged with the virus
[b] QD 1-7 started on the day preceding the challenge
[c] Mean value (20 mice)

to the corresponding nucleotide level. Neplanocin A, which ranked among the most potent inhibitors of AdoHcy-ase, reduced viral cytopathogenicity against VSV at the lowest dose level (ID_{50} 0.01 μg/ml) when compared with other AdoHcy-ase inhibitors (17). However, despite its superior potency with respect to 3-deaza-C-Ado (the carbocyclic analogue of 3-deazaadenosine), neplanocin A resulted more cytotoxic (16). In various cell lines the metabolic conversion of neplanocin A to the nucleotide triphosphate and the subsequent transformation of the latter into the AdoMet analogue, S-neplanocylmethionine, has been demonstrated (21). Moreover, the conversion of neplanocin A to these metabolites has been correlated to the inhibition of RNA synthesis and cytotoxicity (22). Because the antiviral effects of neplanocin A appeared to be mediated solely by its inhibition of AdoHcy-ase, blockage of its anabolism to the nucleotide level was considered to be an attractive strategy to achieve higher antiviral selectivity. To this end, based on the knowledge that 3-deazaadenosine nucleosides are very poor substrates for adenosine kinase (18) (the first enzyme in the activation pathway to the triphosphate), the synthesis of 3-deazaneplanocin A was proposed (23,24). This new member of the carbocyclic nucleoside family retained the potent inhibition of AdoHcy-ase to even higher levels than those achieved with neplanocin A (23,24), and as anticipated, it showed a more selective antiviral effect both _in vitro_ and _in vivo_ (vide supra). In this regard it is interesting to note that neplanocin A failed to show any appreciable antiviral effect _in vivo_ where it proved to be extremely cytotoxic and lethal to mice (16).

2′,3′-Dideoxy-2′-fluoro-ara-A (**13**, ddF-ara-A) and 2′,3′-dideoxy-2′-fluoro-ara-I (**14**, ddF-ara-I).

Chemistry. The chemical rationale for the synthesis of these compounds was based on the need to overcome the extreme acid instability of the parent dideoxyadenosine (ddA) and dideoxyinosine (ddI) nucleosides. Both ddA and ddI have half lives of approximately half a minute at pH 1 (37°C) which makes them unsuitable for oral use (25). Based on the mechanism of acid-catalyzed cleavage of purine nucleosides (26), it was reasoned that an electronegative substituent at the 2′-position, such as fluorine, would destabilize the resulting oxonium ion and thus increase the stability of the glycosylic bond (Figure 2). Fluorine was also considered an ideal isostere for hydrogen because of its relative small size. The only remaining issue was the selection of the stereochemical orientation of the fluorine substituent (α or β) that was required to maintain the anti-HIV activity of the parent dideoxynucleosides. Both fluorine-substituted isomers of ddA were synthesized and shown to be hydrolytically very stable (no decomposition detected after 24 h at pH 1/37° C). However, only the isomer with the β-fluorine atom (ddF-ara-A) retained the potent anti-HIV activity of the parent ddA (25).

The inactive α-fluorine isomer (ddF-A, **8**, Scheme 3) was obtained in four steps from 3′-deoxy-ara-A (27). The procedure involved protection of the the 5′-hydroxyl group with dimethoxytrityl chloride to give **7**, activation of the 2′-hydroxyl group via formation of the corresponding triflate derivative, inversion of configuration at the 2′-position by S_N2 displacement using tetra-_n_-butylammonium fluoride, and removal of the dimethoxytrityl group with dichloroacetic acid. A similar displacement reaction strategy employing cordycepin (3′-deoxyadenosine) failed in our hands to produce **15**. Instead, the elimination product was isolated. Recently, however, this type of displacement has been realized by Herdewijn et al., albeit in very low yield (28).

Two different synthetic approaches, both converging on the important known intermediate (6-amino-9-β-_D_-2-deoxy-2-fluroarabinofuranosyl)-9_H_-purine (**11**), were developed for the synthesis of the active compounds ddF-ara-A and ddF-ara-I. This intermediate, originally synthesized by Fox and coworkers (29), was prepared using the improved, general procedure of Montgomery et al (30), which involved condensation of 6-chloropurine with 3-_O_-acetyl-5-

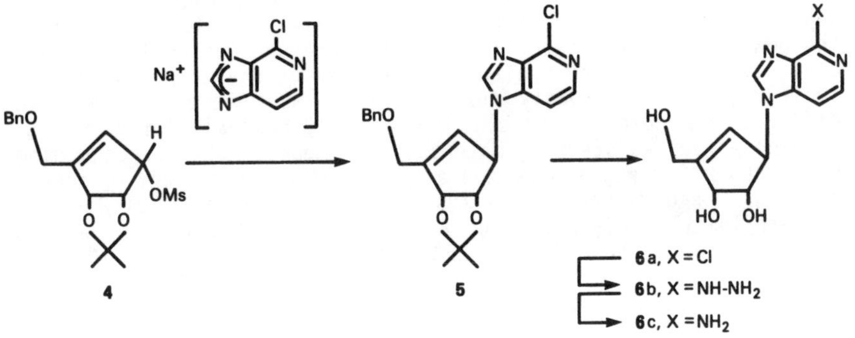

BnO Na+ Cl N N N H OMs O O 4 BnO Cl N N N O O 5 HO X N N N HO OH

6a, X = Cl

6b, X = NH-NH$_2$

6c, X = NH$_2$

Scheme 2. Synthesis of 3-deazaneplanocin A.

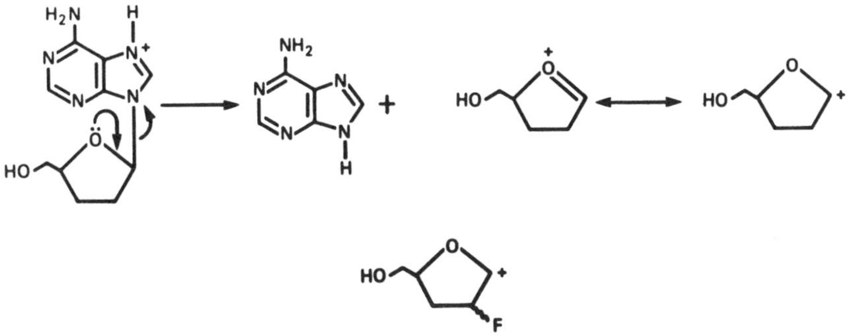

H_2N H N N+ N N HO O NH$_2$ N N N N H + HO O+ HO O + HO O + F

Figure 2. Mechanism of acid hydrolysis of dideoxynucleosides and acid stabilization rationale.

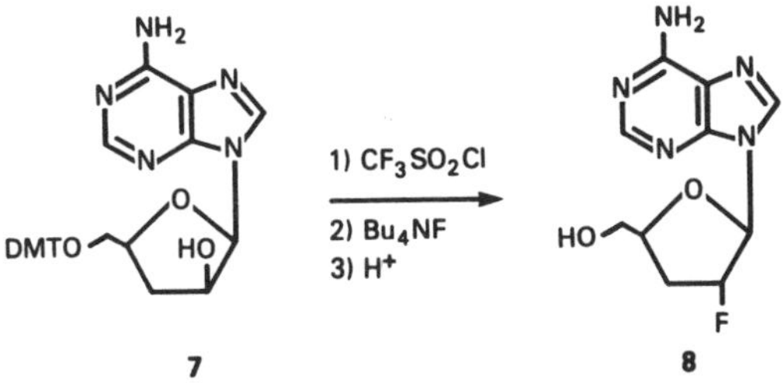

NH$_2$ N N N N DMTO O HO 7

1) CF_3SO_2Cl
2) Bu_4NF
3) H^+

NH$_2$ N N N N HO O F 8

Scheme 3. Synthesis of 2′,3′-dideoxy-2′-fluoro-ribo-A.

O-benzoyl-2-deoxy-2-fluoro-D-arabinofuranosyl bromide (25). The required, functionalized halosugar was prepared in essentially the same manner as described by Fox et al. (31) and, as expected, four isomers were obtained from the condensation reaction. After separation and characterization of the correct 6-chloro isomer, the required intermediate (**11**) was obtained upon ammonolysis. All chemical, optical, and spectral properties of the compound matched those reported previously for **11** (25). More recently, we have employed the more accessible 1,3,5-tri-(O-benzoyl)-2-fluoro-α-D-arabinose (**9a**) as starting material (Scheme 4). The corresponding bromo sugar **9b**, readily formed after treatment with HBr in acetic acid, was used directly in a fusion reaction (neat 85°C) with the trimethylsilyl (TMS)-protected 6-chloropurine to give the desired condensation product **10** in 25% yield. Treatment of **10** with methanolic ammonia under pressure produced the identical intermediate **11**. Selective protection of the primary alcohol function of **11**, accomplished with tert-butyldimethylsilyl chloride, gave compound **12**, and treatment of this compound with phenyl chlorothionocarbonate/DMAP, afforded the corresponding 3′-O-phenoxythionocarbonate **13**. Reductive deoxygenation to **14** with tri-n-butyltin hydride/AIBN in refluxing toluene, and removal of the tert-butyldimethylsilyl group with tetra-n-butylammonium fluoride in THF, afforded ddF-ara-A (**15**). The corresponding inosine analogue ddF-ara-I (**16**) was prepared from ddF-ara-A either by enzymatic (adenosine deaminase) or chemical deamination ($NaNO_2$-/HOAc).

Antiviral Activity. In ATH8 cells infected with the HIV virus, both ddF-ara-A (25) and ddF-ara-I were approximately as active and potent as AZT, ddA, and ddI, in protecting the cells against the cytopathic effects of the virus. As shown in Figure 3, ddF-ara-I afforded complete protection at 10 μM concentration and beyond. Significant anti-HIV activity with ddF-ara-A was also reported in infected MT-4 cells; however, the compound proved to be inferior to its parent ddA in such test system (31). As it is the case with ddA, adenosine deaminase readily transformed ddF-ara-A to ddF-ara-I without detriment to the measured anti-HIV activity. It was therefore important to investigate the effects of the fluorine substituent on a second degradative enzyme, purine nucleoside phosphorylase, which is capable of rapidly cleaving the glycosylic linkage of ddI and rendering the compound inactive. Consistent with its increased chemical stability towards acid-catalyzed cleavage of the glycosylic bond, ddF-ara-I was shown to be totally inert towards the action of purine nucleoside phosphorylase (32).

In order to study the effect of the fluorine substituent on the gastric stability and oral bioavailability of the active ddF-ara-A, the NCI conducted comparative studies in dogs treated separately with ddA or ddF-ara-A. Under the same conditions, without any buffering, ddF-ara-A showed superior bioavailability (87%) compared to that of ddA (30%) (33). The oral bioavailability of ddF-ara-A was even greater than that obtained with buffered (61%), or enteric coated (67%) ddA (33).

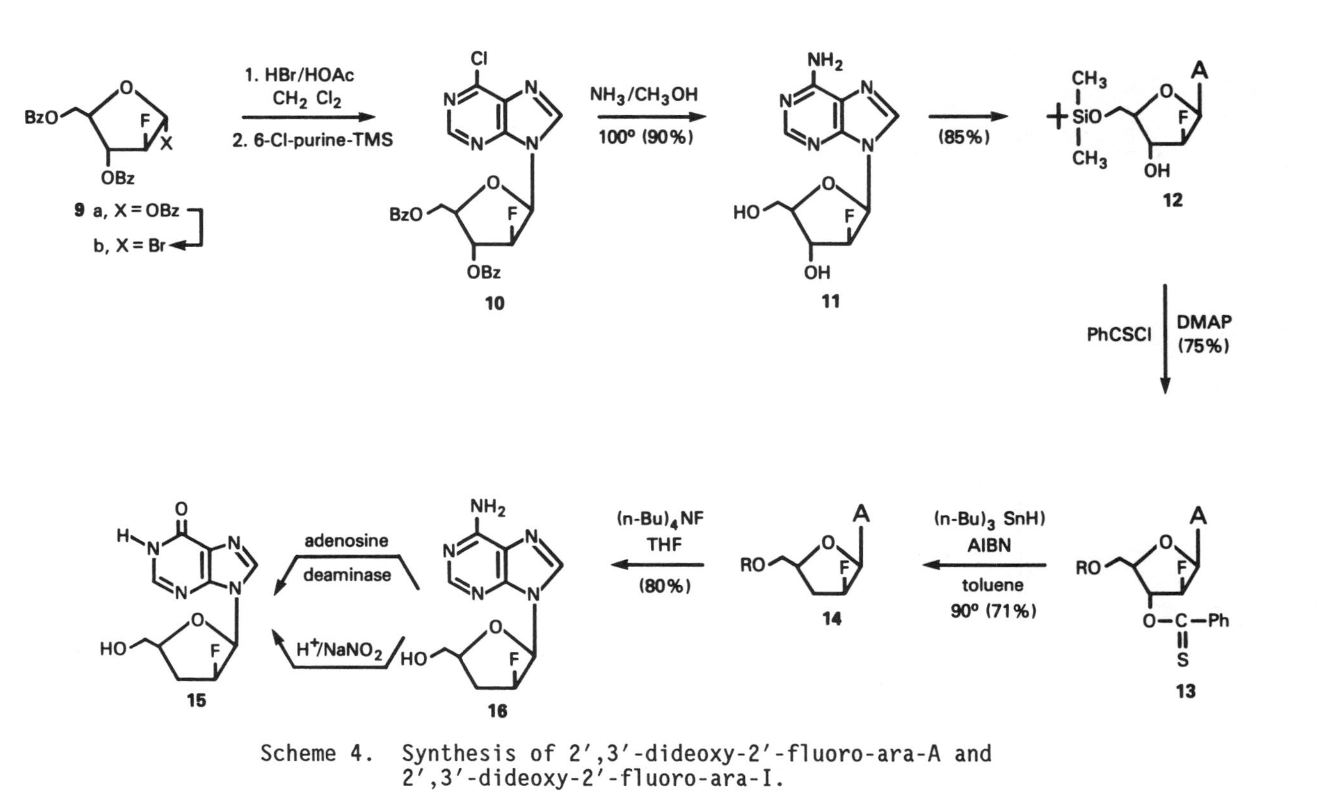

Scheme 4. Synthesis of 2',3'-dideoxy-2'-fluoro-ara-A and 2',3'-dideoxy-2'-fluoro-ara-I.

Mechanism of Action. Although no direct mechanistic studies as yet have been performed with either ddF-ara-A or ddF-ara-I, it is expected that both compounds will be anabolized to the corresponding nucleotide 5′-triphosphate analogues in a similar fashion as their parent dideoxynucleosides (ddA and ddI) (34). Dideoxynucleotide triphosphate analogues are known to function as effective inhibitors of HIV reverse transcriptase that block DNA synthesis by chain termination (35). Metabolically it is possible, as demonstrated for ddI, that ddF-ara-IMP will be converted to ddF-ara-AMP via the same efficient pathway that converts IMP to AMP (34).

9-(2-Deoxy-2-dihydroxyphosphinylmethyl-β-D-ribofuranosyl)adenine (26).

Chemistry. Compound 26 corresponds to units one and two of a proposed modified 2′-5′-linked oligoadenylate trimer "core" (**17**, Figure 4), which due to is hydrolytically resistant carbon-phosphorous linkage, is expected to be stable towards degradation by the interferon (IF)-induced 2′-phoshodiesterase (2′-PD). Previous studies of synthetic analogues of the trimer "core", A2′p5′A2′p5′A (lacking the triphosphate group at the 5′-end), which included modifications on the sugar moiety designed to make the molecules more stable towards 2′-PD, showed significantly better antiproliferative and antiviral activity than the unstable parent "core" (36). However, it was later demonstrated that all of the biological activities of the so called "stable cores" were due to the monomeric components which were released from the trimers by nonspecific serum enzymes (36b,e). Thus, the search for stable analogues which would not yield active antimetabolites upon nonspecific hydrolysis continues (37). The first objective of this project, which was the synthesis and biological testing of the phosphonate monomer **26**, has been achieved, and the assembly of the required trimer **17** is in progress.

We have accomplished the synthesis of **26** as shown in Scheme 5. Starting with the versatile intermediate 1,2:5,6-di-O-isopropylidene-α-D-glucofuranose, the phosphonate **18** was prepared essentially in the same manner as reported by Moffatt and coworkers (38). Selective hydrolysis of the 5,6-O-isopropylidene function, followed by protection of the resulting primary alcoholic group of the diol with the benzoyl group, afforded compound **20**. Removal of the 1,-2-O-isopropylidene group gave the intermediate pyranoside **21**, which was subjected to periodate oxidation to give the desired phosphonate furanoside **22**. As reported in similar ring contractions, the excised carbon atom ends up as a 3′-O-formyl moiety (29). Selective removal of this formate group, and reacylation of both the 1′- and the 3′-hydroxyls with acetic anhydride, set the stage for the condensation of the phosphonate sugar **24** with persilylated 6-chloropurine, which was performed under Lewis acid catalysis (trimethylsilyl triflate). The condensation reaction gave a remarkably good yield (82%) of the desired β-isomer **25**, which upon treatment with saturated methanolic ammonia, gave the desired target compound **26**. The structural assignment of **26** rests on solid evidence from UV, CD, and NOE experiments as illustrated in Figure 5.

Antiviral Activity. The monomeric structure **26** showed neither antitumor nor antiviral activity. Upon *in vitro* evaluation against herpes simplex virus type 1 (HSV-1) and influenza virus type A_2, no evidence of inhibition of viral cytopathogenicity was detected at doses exceeding 320 μg/ml (39).

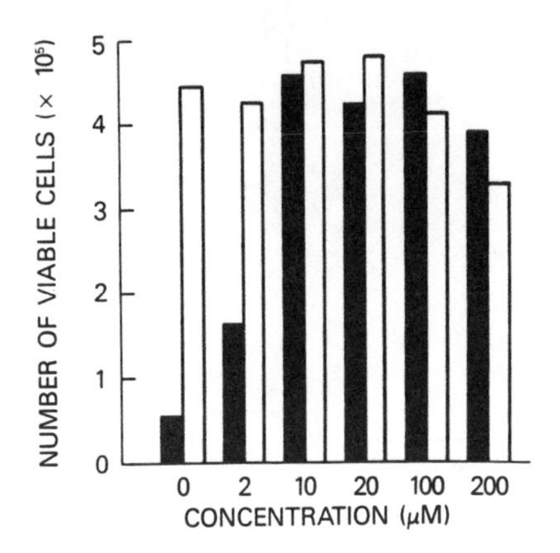

Figure 3. Inhibition of the cytopathic effect of HIV by ddF-ara-I.

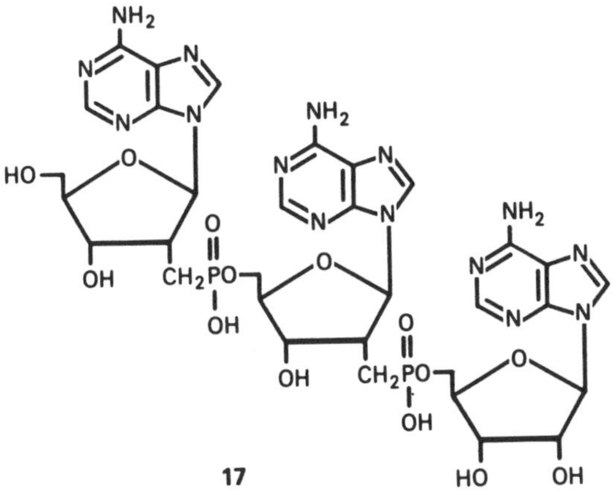

Figure 4. Chemical structure of the oligoadenylate phosphonate trimer.

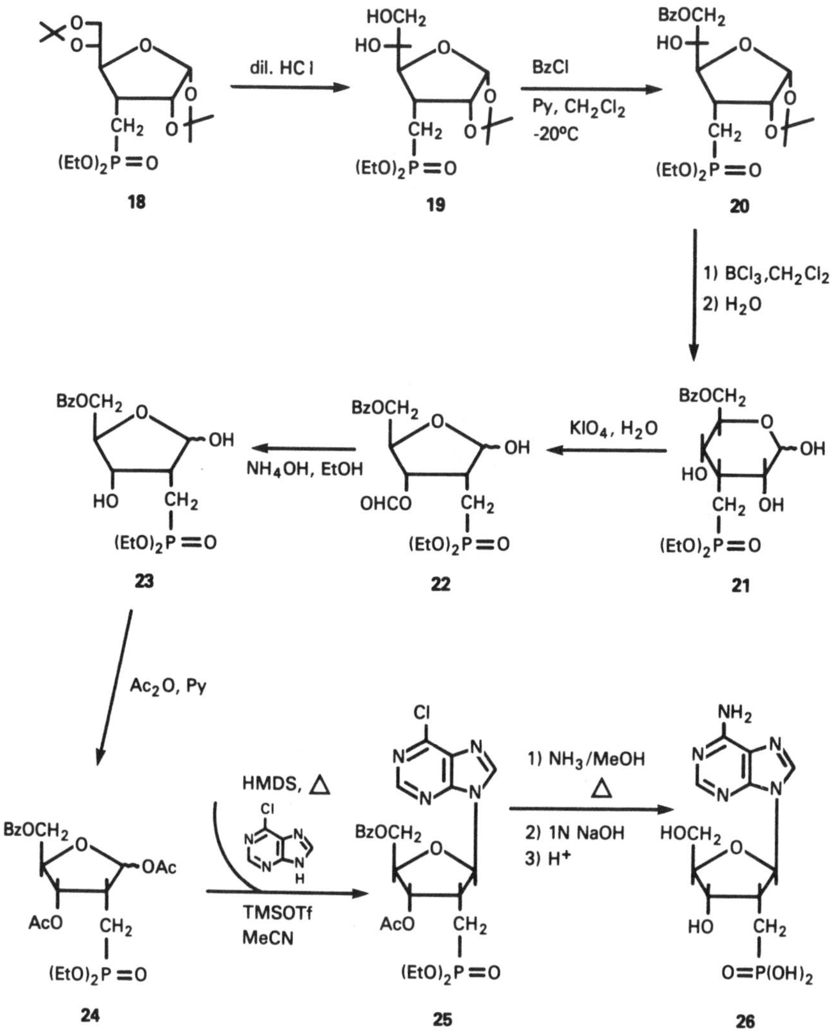

dil. HCl

BzCl

Py, CH_2Cl_2

-20°C

1) BCl_3, CH_2Cl_2

2) H_2O

KIO_4, H_2O

NH_4OH, EtOH

Ac_2O, Py

HMDS, Δ

TMSOTf

MeCN

1) NH_3/MeOH Δ

2) 1N NaOH

3) H^+

18 19 20 21 22 23 24 25 26

Scheme 5. Synthesis of 9-(2-deoxy-2-dihydroxyphosphinylmethyl)-β-D-ribofuranosyl)adenine

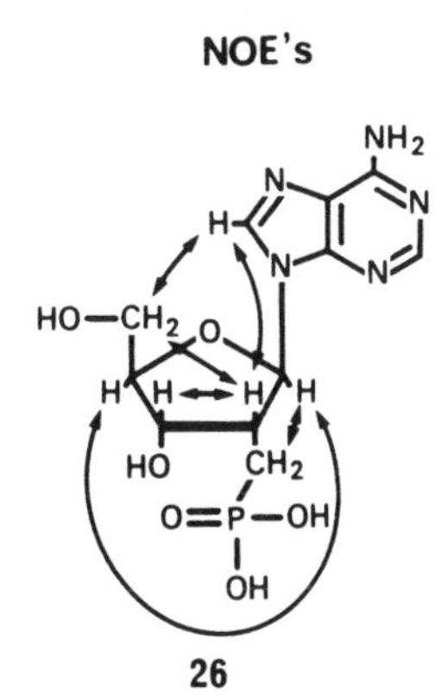

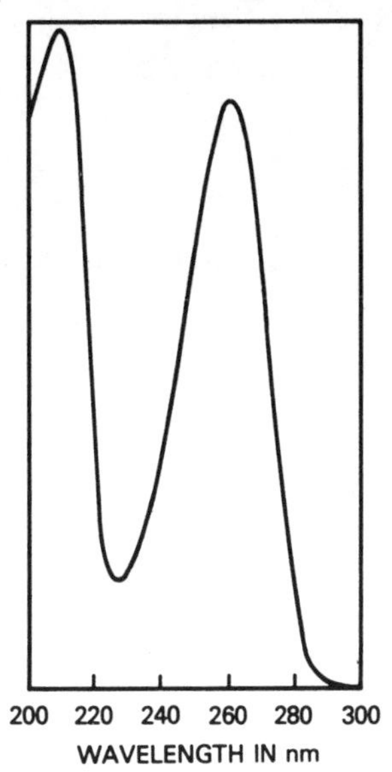

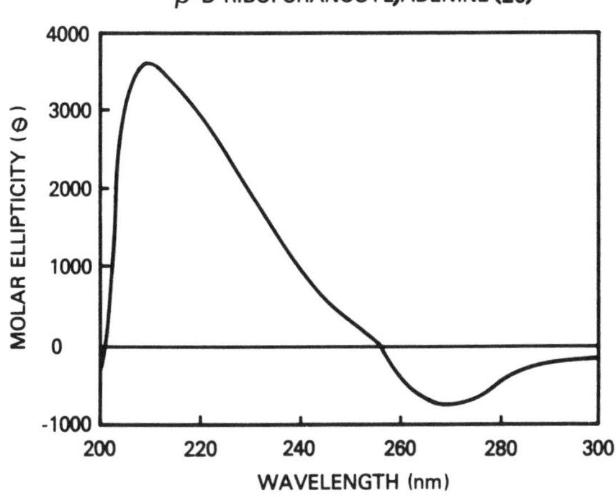

Figure 5. Spectral data in support of structure **26**.

Literature Cited

1. Robins, R.K.; Revankar, G.; McKernan, P.A.; Murray, B.K.; Kirsi, J.J.; North, J.A. Adv. Enzyme Res. 1985, 24, 29.
2. Lin, J-C.; Pagano, J.S. Pharmac. Ther. 1985, 28, 135.
3. Prisbe, E.J.; Martin, J.C.; McGee, D.P.C.; Barker, M.F.; Smee, D.F.; Duke, A.E.; Matthews, T.R.; Verheyden, J.P.H. J. Med. Chem. 1986, 29, 671.
4. De Clercq, E. Antimicrob. Agents Chemother. 1985, 28, 84.
5. Lim, M-I.; Moyer, J.D.; Cysyk, R.L.; Marquez, V.E. J. Med. Chem. 1984, 27, 1536.
6. Marquez, V.E.; Lim, M-I.; Treanor, S.P.; Plowman, J.; Priest, M.A.; Markovac, A.; Khan, M.S.; Kaskar, B.; Driscoll, J.S. J. Med. Chem. 1988, 31, 1687.
7. Arita, M.; Adachi, K.; Ohno, M. Nucleic Acid Res., Symp. Ser. 1983, 12, 25.
8. Arita, M.; Okumoto, T.; Saito, T.; Hoshino, Y.; Fukukawa, K.; Shuto, S.; Tsujino, M.; Sakakibara, H.; Ohno, M. Carbohydr. Res. 1987, 171, 233.
9. Kim, S.K.; Marquez, V.E.; Driscoll, J.S. Laboratory of Medicinal Chemistry, NCI, NIH. Unpublished results.
10. Kim, S.K.; Marquez, V.E. 194th National Meeting of the American Chemical Society, New Orleans, Louisiana, August 30-September 4, 1987, CARB 6.
11. De Clercq, E.; Beres, J.; Benrude, W.G. Mol. Pharmacol. 1987, 32, 286.
12. Marquez, V.E.; Driscoll, J.S., Laboratory of Medicinal Chemistry, NCI, NIH. Unpublished results.
13. Kang, G.J.; Cooney, D.A.; Moyer, J.D.; Kelley, J.A.; Kim, H-Y.; Marquez, V.E.; Johns, D.G. J. Biol. Chem. 1988, in press.
14. Tseng, C.K.H.; Marquez, V.E. Tetrahedron Lett. 1985, 26, 3669.
15. X-ray crystallographic analysis for 3-deazaneplanocin A was performed by Dr. Barry M. Goldstein, Department of Biophysics, Medical Center, University of Rochester, Rochester, N.Y.
16. De Clercq, E. Antimicrob. Agents Chemother. 1985, 28, 84.
17. De Clercq, E.; Cools, M. Biochem. Biophys. Res. Commun. 1985, 129, 306.
18. Montgomery, J.A.; Clayton, S.J.; Thomas, H.J.; Shannon, W.M.; Arnett, G.; Bodner, A.J.; Kion, I.; Cantoni, G.L.; Chiang, P.K. J. Med. Chem. 1982, 25, 626.
19. De Clercq, E.; Bergstrom, D.E.; Holy, A.; Montgomery, J.A. Antiviral Res. 1984, 4, 119.
20. Marquez, V.E.; Lim, M-I. Med. Res. Rev. 1986, 6, 1.
21. Keller, B.T.; Borchardt, R.T. Biochem. Biophys. Res. Commun. 1984, 120, 131.
22. Glazer, R.I.; Knode, M.C. J. Biol. Chem. 1984, 259, 12964.
23. Glazer, R.I.; Hartman, K.D.; Knode, M.C.; Richard, M.M.; Chiang, P.K.; Tseng, C.K.H.; Marquez, V.E. Biochem. Biophys. Res. Commun. 1986, 135, 688.
24. Glazer, R.I.; Knode, M.C.; Tseng, C.K.H.; Haines, D.R.; Marquez, V.E. Biochem. Pharmacol. 1986, 35, 4523.
25. Marquez, V.E.; Tseng, C.K.H.; Kelley, J.A.; Mitsuya, H.; Broder, S.; Roth, J.S.; Driscoll, J.S. Biochem. Pharmacol. 1987, 36, 2719.

26. York, J.L. J. Org. Chem. 1981, 46, 2171.
27. 3'-deoxy-ara-A was synthesized by Dr. Terrence C. Owen (University of South Florida) by the method of Hansske, F. and Robins, M.J. (J. Am. Chem. Soc. 1983, 105, 6736) under contract NO1-CM-437639 to the DS&CB, NCI.
28. Herdewijn, P.; Pauwels, R.; Baba, M.; Balzarini, J.; De Clercq, E. J. Med. Chem. 1987, 30, 2131.
29. Wright, J.A.; Taylor, N.F.; Fox, J.J. J. Org. Chem. 1969, 34, 2632.
30. Montgomery, J.A.; Shortnacy, A.T.; Carson, D.A.; Secrist III, J.A. J. Med. Chem. 1986, 29, 2389.
31. Reichman, U.; Watanabe, K.A.; Fox, J.J. Carbohydr. Res. 1975, 42, 233.
32. Cooney, D.A., Laboratory of Pharmacology, NCI, NIH. Unpublished results.
33. Tomaszewski, J.E.; Grieshaber, C.K., Toxicology Branch, NCI, NIH. Unpublished results.
34. Ahluwalia, G.; Cooney, D.A.; Mitsuya, H.; Fridland, A.; Flora, K.P.; Hao, Z.; Dalal, M.; Broder, S.; Johns, D.G. Biochem. Pharmacol. 1987, 36, 3797.
35. Mitsuya, H.; Broder, S. Proc. Natl. Acad. Sci. USA 1986, 83, 1911.
36. a. Baglioni, C.; D'Alessandro, S.B.; Nilsen, T.S.; den Hartog, J.A.J.; Crea, R.; Van Boom, J.H. J. Biol. Chem. 1981, 256, 3253.
    b. Chapekar, M.S.; Glazer, R.I. Biochem. Biophys. Res. Commun. 1983, 115, 137.
    c. Doetsch, P.; Wu, J.M.; Sawada, Y.; Suhadolnik, R.J. Nature, 1981, 291, 355.
    d. Eppstein, D.A.; Marsh, Y.V.; Schryver, B.B.; Larsen, M.A.; Barnett, J.W.; Verheyden, J.P.H.; Prisbe, E.J. J. Biol. Chem. 1982, 257, 13390.
    e. Eppstein, D.A.; Marsh, Y.V.; Schryver, B.B. Virology, 1983, 131, 341.
    f. Eppstein, D.A.; Van der Pas, M.A.; Schryver, B.B.; Sawai, H.; Lesiak, K.; Imai, J.; Torrence, P.F. J. Biol. Chem. 1985, 260, 3666.
    g. Imai, J.; Johnston, M.I.; Torrence, P.F. J. Biol. Chem. 1982, 257, 12739.
    h. Lee, C.; Suhadolnik, R.J. FEBS Lett. 1983, 157, 205.
    i. Sawai, H.; Imai, J.; Lesiak, K.; Johnston, M.I.; Torrence, P.F. J. Biol. Chem. 1983, 258, 1671.
    j. Suhadolnik, R.J.; Devash, Y.; Reichenbach, N.L.; Flock, M.B.; Wu, J.N. Biochem. Biophys. Res. Commun. 1983, 111 205.
37. Eppstein, D.A.; Schryver, B.B.; Marsh, Y.V. J. Biol. Chem. 1986, 261, 5999.
38. Albrecht, H.P.; Jones, G.H.; Moffatt, J.G. Tetrahedron, 1984, 40, 79.
39. Marquez, V.E.; Tseng, C.K.H., Laboratory of Medicinal Chemistry, NCI, NIH. Unpublished results.

RECEIVED January 25, 1989

Chapter 11

Nucleotide Dimers as Anti Human Immunodeficiency Virus Agents

Elliot F. Hahn[1], Mariano Busso[2], Abdul M. Mian[3], and Lionel Resnick[2]

[1]IVAX Corporation, 8800 NW 36th Street, Miami, FL 33178
[2]Departments of Dermatology and Pathology, Mount Sinai Medical Center, 4300 Alton Road, Miami Beach, FL 33140
[3]Department of Oncology, University of Miami Medical School, Miami, FL 33131

A series of nucleotide homo and heterdimers were synthesized from nucleosides which include azidothymidine, 2', 3'-dideoxyadenosine and 2'3'-dideoxyinosine. The compounds were evaluated with respect to anti-HIV activity, cytotoxicity, in vitro stability and in vivo distribution. On an equimolar basis, greater anti-HIV potency and enhanced cytotherapeutic indices were obtained with heterodimers relative to the corresponding monomers. In vitro stability of the dimer phosphate linkage was found to be species dependent. After intravenous administration to rats, the distribution of 3'-azido-3'-3'-deoxythymidilyl-(5'5')-2',3'-dideoxy;-5'-adenylic acid, 2-cyanoethyl ester (AZT-P(CyE)ddA) and AZT in plasma and brain was similar.

In 1981, the acquired immunodeficiency syndrome (AIDS) was reported as a new clinical entity (1-3). An intensive research effort led to the identification of the etiologic agent responsible for the disease. (4-7). The pathogen, human immunodeficiency virus (HIV), is a non-oncogenic retrovirus closely associated with the lentivirus family (8,9).

The discovery of HIV led to the development of therapeutic strategies directed against the virus. The different stages in the replicative cycle of HIV provide various targets at which antiviral agents may intervene.

0097–6156/89/0401–0156$06.00/0

These have been the subject of a number of reviews (10,11) and include: (1) the binding stage at which CD4 receptor decoys and dextran sulfates act, (2) entry into the target cell where drugs that block fusion or uncoating might intervene, (3) transcription of RNA to DNA, the level at which reverse transcriptase inhibitors would function, (4) integration of DNA and expression of viral genes, where "integrase" inhibitors or "anti-sense" construct would act, (5) viral protein production and assembly could be modified by protease inhibitors and agents which affect, myristylation or glycosylation and finally, (6) budding of the virus which may be prevented by interferons.

The most effective approach has involved the synthesis of inhibitors of reverse transcriptase, the unique enzyme associated with the replication of HIV. The rationale for the use of antiviral agents in HIV disease is predicated on the assumption that replication is necessary for the development of progressive disease. Inhibition of HIV may permit the regeneration or prevent additional deterioration of the immune system. These drugs would provide the greatest possibility of obtaining an immediate clinical impact on the course of the disease.

Since the discovery of reverse transcriptase occurred in 1970 (12,13), compounds that inhibit the enzyme had been developed and were available for evaluation at the advent of the AIDS crisis. To date, the most potent and selective anti-HIV compounds are a series of 2',3'-dideoxynucleoside analogs. These compounds are thought to be successively phosphorylated by host cell enzymes to yield 2',3'-dideoxynucleoside-5'-triphosphates, which are analogs of 2'-deoxynucleoside-5'-triphosphates, the natural substrates for cellular DNA polymerases and reverse transcriptase. The 2',3'-dideoxynucleoside-5'-triphosphates function as substrates for HIV reverse transcriptase and terminate viral DNA chain elongation by incorporation into the viral genome (14). Only 3'-azido-2',3'-dideoxythymidine (AZT) has been approved by the FDA for the treatment of HIV infection. AZT was chosen for clinical evaluation on the basis of its selective *in vitro* antiviral effect against HIV. The clinical trials with AZT have focused on patients with AIDS and AIDS-related complex (ARC). In these patients, AZT induced clinical and laboratory improvements. AZT therapy also exhibits a degree of success in reversing HIV induced dementia (15,16,17). Current studies aim to determine if AZT is effective in preventing the development of AIDS in asymptomatic HIV seropositive individuals. AZT therapy was associated with toxicities that may limit its use, which primarily involved bone marrow suppression in the form of anemia and neutropenia (15,16). Therefore, other strategies which involve combinations of anti HIV agents are also being pursued. The most promising of the 2',3'-

dideoxynucleosides to be used individually or in combination with AZT are 2',3'-dideoxycytidine (ddC), and 2',3'-dideoxyadenosine (ddA). The ultimate success of these agents against HIV is determined in the clinical arena. Combinations of drugs which permit a reduction in individual doses would be a means of decreasing toxicity.

Another approach to combination therapy involves the use of either homo or heterodimers of specific dideoxynucleosides. We hypothesized that when the dideoxynucleosides are linked via a phosphate bridge the dimerized compounds could provide superior pharmacological effects for the following reasons 1) two different nucleosides are delivered to the cell simultaneously, 2) a masked phosphate is present with masking unit also being active, 3) the agent could function as a prodrug, and 4) activity as the intact dimer was possible. In general, the initial phosphorylation of the nucleosides to yield mononucleotides is a limiting step for the formation of these active metabolites (18). However, AZT is unique and the initial phosphorylation is not rate limiting. If the nucleotide dimers cross the cell membrane (passive diffusion) and are hydrolysed intracellularly (data indicate that at least part of the dimer crosses the cell membrane and is hydrolysed intracellularly - manuscript in preparation), they would yield 1 mole of mononucleotide and nucleoside. It is possible that the nucleotide produced will be further anabolized to its triphosphate level rather than be a substrate for the phosphatases and produce the nucleoside. Therefore, this process will eliminate the need for the initial obligatory phosphorylation step and may result in a superior therapeutic index. If the nucleotide dimers are hydrolysed before their uptake, the nucleotide phosphatases will convert the mononucleotide into the nucleoside. In this fashion, the dimers may act as "depot forms" for their respective nucleosides and may yield a favorable therapeutic index.

We have synthesized a series of compounds which are homodimers or heterodimers of specific dideoxynucleosides and have shown that these agents possess activity against HIV. The phosphate linked dimers of the dideoxynucleosides have the general formula shown in Fig. 1 where R^1 and R^2 may be AZT, ddA or ddI and R^3 may be hydrogen, cyanoethyl or either a metal anion or organic anion salt. It should be noted that R^1 and R^2 may be derived from the same or different dideoxynucleosides to generate phosphate-linked compounds that are either homo or hetero dimers. The dimers may be prepared by coupling the nucleoside to be esterified with a nucleoside 5'-phosphate in the presence of a arylsulfonyl condensing agent and a base such as imidazole (Fig. 2). A general procedure for the synthesis of AZT-P-ddA, which is applicable for the synthesis of all nucleotide dimers is as follows: The nucleoside-5'-cyanoethyl phosphate was synthesized using a modification of the procedure described by Tener (19). Thus, the barium salt of cyanoethyl phosphate (5.0g) was converted to the pyridinium salt under anhydrous conditions and reacted with AZT (2.0g) in the presence of dicyclohexyl-carbodiimide (6.18g) at room temperature for 48 hours. The reaction was stopped by addition of water (10ml) and the product was purified by column chromatography to obtain 2.07g (69%) of the desired product. A solution of AZT 5'-cyanoethyl phosphate (600mg, 1.5mmol) and 2',3'-dideoxyadenosine (352mg, 1.5mmol) was prepared in 60 ml of distilled pyridine. To this solution, 1.21 g (4mmol) of 2,4,6-triisopropylbenzenesulfonyl chloride was added and stirred for 30 min, followed by the addition of 0.984 ml (12 mmol) of N-methyl-imidazole. After stirring at room temperature for 18 hours, the solvent was removed under vacuum and the residue was purified by column chromatography using silica gel. The cyanoethyl phosphate ester was hydrolyzed at room temperature with 60 ml of 15% ammonium hydroxide solution and the product was purified by flash chromatography using silica gel to obtain AZT-P-ddA in 73% yield.

To obtain the cyanoethyl phosphate ester, the crude product prior to hydrolysis with ammonium hydroxide was purified by chromatography.

In initial studies, we determined the anti-HIV activity of the dimers using assays which measured inhibition of syncytium formation, reverse transcriptase production and HIV antigen expression.

Anti-HIV Evaluation:

A syncytium inhibition assay that is a safe, simple, rapid, quantitative and sensitive screening system has been developed to detect potential anti-HIV drugs. The assay has been standardized and validated for high capacity anti-HIV screening. Compounds can be identified and prioritized based upon anti-HIV effects. Potency is expressed as effective dose 50%(ED50), toxicity as inhibitory dose 50% (ID50) and the cytotherapeutic index as ID50/ED50.

1) Inhibition of syncytium formation. MT-2 cells are used as targets because of their sensitivity to HIV infection and the formation of giant syncytia that are quantifiable. The number and the time necessary for the production of syncytia is a function of the input virus inoculum. Target MT-2 cells are exposed to DEAE-Dextran (25ug/ml, Sigma) for 20 minutes, washed, and infected with

HIV-1(III-6) (MOI = 0.001) at 37°C in humidified air containing 5% CO_2. After 96 hours syncytia are counted in a microtiter configuration and compared to controls. Uninfected and infected MT-2 cells without exposure to drug and uninfected cells exposed to drugs are used as controls. All cultures are performed in triplicate on two sets of experiments.

To investigate the inhibitory effects of the drugs on HIV-induced syncytia formation (20), the nucleotide dimers were compared to their monomers and their combinations at multiple concentrations (Fig. 3). AZT-P-ddA and AZT-P(CyE)-ddA exerted the strongest protective effect against the development of HIV-induced syncytia. AZT-P-ddA and its cyanoethyl phosphate derivative at a concentration of 0.5um completely protected MT-2 cells from the formation of syncytia. AZT required a concentration of 1uM, ddA, 10uM, and the combination of AZT + ddA required 0.5uM to achieve full protection. No anti-HIV inhibitory effects were seen at concentrations below 0.01uM.

2) **Inhibitory effect on reverse transcriptase production**.

Reverse transcriptase assays were performed as previously described (20). When MT-2 cells were infected by HIV, peak reverse transcriptase levels in the control cultures without drug were greater than 50,000 CPM/ml (uninfected control cultures gave background counts of 300cpm/ml) (Fig. 4). A significant inhibition of HIV reverse transcriptase production was observed in a dose-dependent manner when HIV-infected MT-2 cells were cultured in the presence of nucleosides and nucleotide dimers. AZT-P-ddA, AZT-P(CyE)-ddA and the combination of AZT + ddA, completely inhibited the products of reverse transcriptase at concentrations >1uM. In comparison to controls, reverse transcriptase production was partially inhibited when these compounds were tested at levels >0.1uM.

The detection of HIV from culture supernatants (infectious viral yield) correlated with detectable levels of reverse transcriptase and HIV-induced syncytia formation. The nucleotide dimers exhibited a higher degree of inhibition for the detection of infectious viral yield when compared to the monomers.

3) **Inhibition of HIV antigen expression and cytopathic effect**.

After fourteen days of infection, 84% of MT-2 cells expressed HIV p24 antigen as detected by indirect immunofluorescence. At a final concentration of 1uM of AZT-P-ddA, AZT-P(CyE)-ddA, or the combination of AZT + ddA, a >70% inhibition of viral antigen expression was

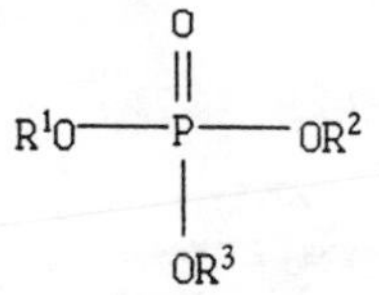

$$R^1O-\overset{\displaystyle O}{\overset{\|}{\underset{\underset{\displaystyle OR^3}{|}}{P}}}-OR^2$$

Figure 1. Structure of Dimers.

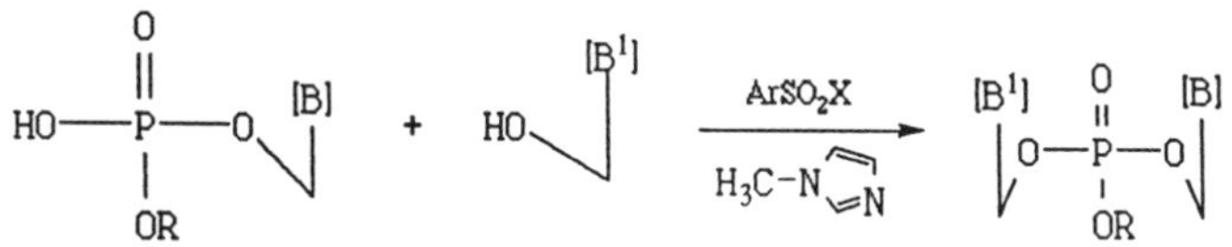

$$HO-\overset{\displaystyle O}{\overset{\|}{\underset{\underset{\displaystyle OR}{|}}{P}}}-O\text{[B]} \;+\; HO\text{[B}^1\text{]} \xrightarrow[H_3C-N\text{(imidazole)}]{ArSO_2X} \text{[B}^1\text{]}O-\overset{\displaystyle O}{\overset{\|}{\underset{\underset{\displaystyle OR}{|}}{P}}}-O\text{[B]}$$

Figure 2. Coupling Scheme for Dimer Synthesis.

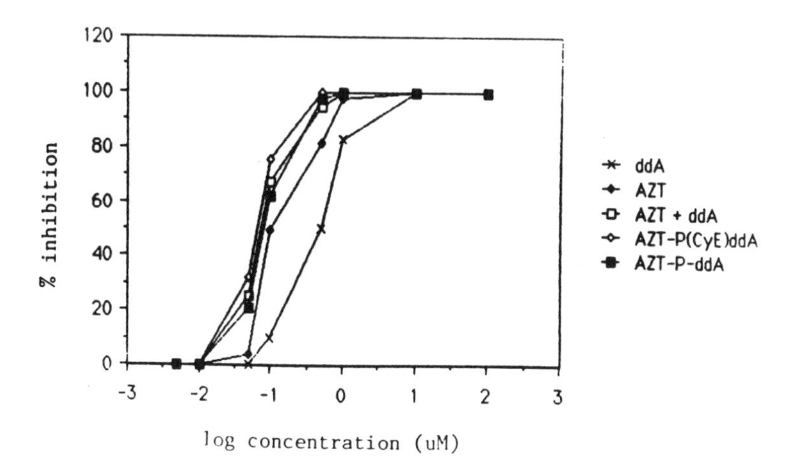

Figure 3. Syncytium Inhibition. The number of syncytia was determined 96 hours after exposure of HIV infected MT-2 cells to drug concentrations (100uM, 10uM, 1uM, 0.5uM, 0.1uM, 0.05uM, 0.01uM and 0.005uM). Each value represents the arithmetic mean of triplicate cultures from two sets of experiments. The mean syncytium number in infected cultures without drug was 421 ± 27.

achieved. AZT (1uM) or ddA (1uM) alone exhibited no inhibition. Similar results were obtained when assessing the inhibition of cytopathic effect by these compounds (Fig. 5).

4) **Cytotherapeutic evaluation**.

The comparative HIV inhibitory effects of nucleosides and nucleotide dimers are shown on Table 1. Using a 14-day assay, monitoring cell viability and the expression of HIV antigen by cellular fluorescence, studies were performed to determine the potency and toxicity of the compounds. Linear regression analysis was performed to determine the ID50 and ED50 for each compound.

The growth inhibitory effects of the compounds on MT-2 cells which were not exposed to the virus, were compared. According to their ID50, the compounds could be classified into three major groups. The compounds with the highest toxicity were AZT-P-AZT, AZT + ddA, AZT-P(CyE), and AZT. Compounds with moderate toxicity were AZT-P-ddA, AZT-P-ddI, and AZT-P(CyE)-ddA. The compounds with the least toxicity were ddI, ddA, and ddA-P(CyE). When the cytotoxic effects of the compounds were tested against the human cell lines, H9 and U937, similar toxic profiles were seen.

The anti-HIV activity of the compounds according to their ED50 revealed two major profiles. The most potent compounds were AZT + ddA, AZT-P-ddA, AZT-P(CyE)-ddA, AZT-P-ddI, and AZT-P-AZT. AZT, ddA, ddI, ddA-P(CyE), and AZT-P(CyE) exhibited weaker activities.

However, the cytotherapeutic indices of AZT-P-ddA, AZT-P-ddI, and AZT-P(CyE)-ddA were the highest.

Based on the data obtained from these studies we focused on further evaluation of AZT-P-ddA and AZT-P(CyE)-ddA. Figure 6 shows the results of incubation of AZT-P-ddA in human plasma at various temperatures. The compound is metabolized at a rate of approximately 10% per hour. This value appears to be species dependent as is seen in Fig. 7. Analysis of the metabolites in human plasma (Fig. 8) shows that the principal compounds formed are AZT and ddI. A similar study of the stability of AZT-P(CyE)-ddA (Fig. 9) at 37° shows that it is metabolized more extensively over 3 hour assay period when compared to AZT-P-ddA. Figure 10 shows that the nature of the products formed is also species related. In human plasma, the major metabolite is AZT-P-ddA which is a result of hydrolysis of the triester.

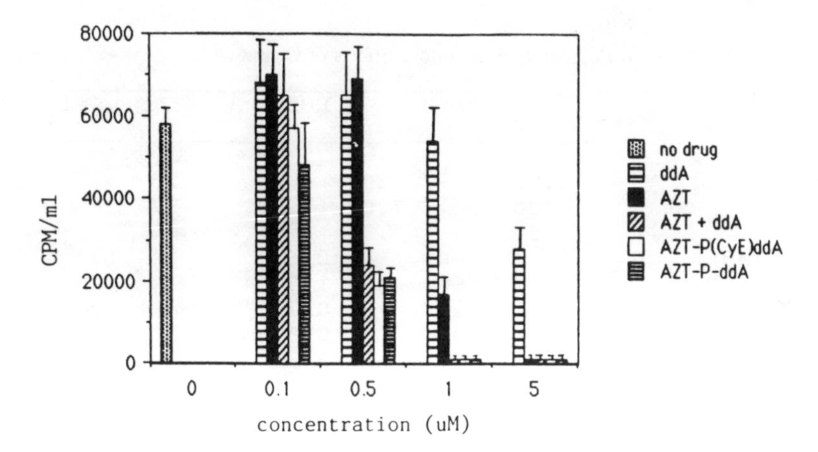

Figure 4. Inhibition of Reverse Transcriptase Activity. Inhibition of reverse transcriptase activity from culture supernatants was determined on day 8. Each value represents the arithmetic mean of triplicate cultures from two sets of experiments. Detectable levels in reverse transcriptase activity occur when CPM/ml are greater than 5,000 (Lines, SD).

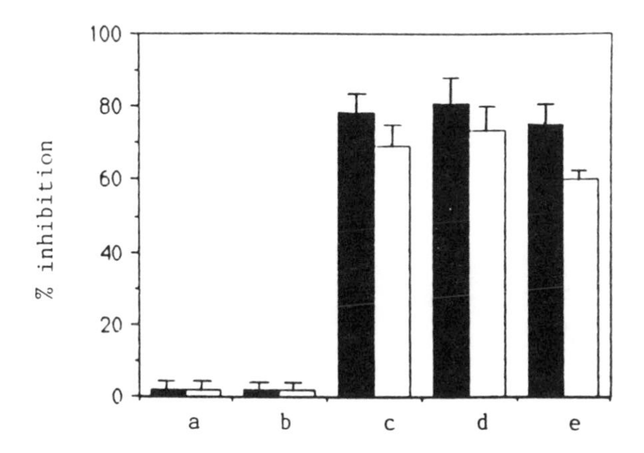

Figure 5. Assessment of HIV Inhibition by Fluorescent and Viable Cell Count. The inhibitory effect of 1uM of drug on HIV expression was assessed on day 14 by indirect immunofluorescence (solid columns) and viable cell numbers (open columns): (a) ddA, (b) AZT, (c) AZT + ddA, (d) AZT-P(CyE)-ddA, and (e) AZT-P-ddA. The results are the arithmetic mean of triplicate cultures from two sets of experiments (Lines, SD).

Table 1. Comparative Inhibitory Effects of Compounds

Drug	ID50	ED50	CTI
AZT	100	4.0	25
ddA	400	7.0	57
ddI	450	7.5	60
AZT + ddA	80	0.6	133
AZT-P-AZT	60	1.5	40
AZT-P(CyE)-ddA	210	0.7	300
AZT-P-ddA	200	0.8	250
AZT-P(CyE)	90	3	30
AZT-P-ddI	240	1	240
ddA-P(CyE)	400	5	80

ID50: Drug concentration required to reduce the number of uninfected MT-2 cells by 50% on day 14.

ED50: Drug concentration achieving 50% inhibition of HIV expression assessed by immunofluorescence on day 14.

CTI: Cytotherapeutic index: ID50/ED50.

The results are expressed as the arithmetic mean of triplicate cultures from two sets of experiments. Linear regression analysis was used to determine the ID50 and ED50.

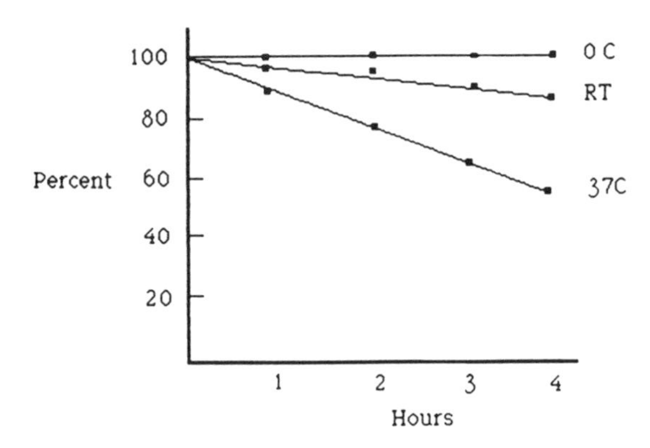

Figure 6. Stability of AZT-P-ddA in Human Plasma (in vitro) at Various Temperatures (4 ug/ml).

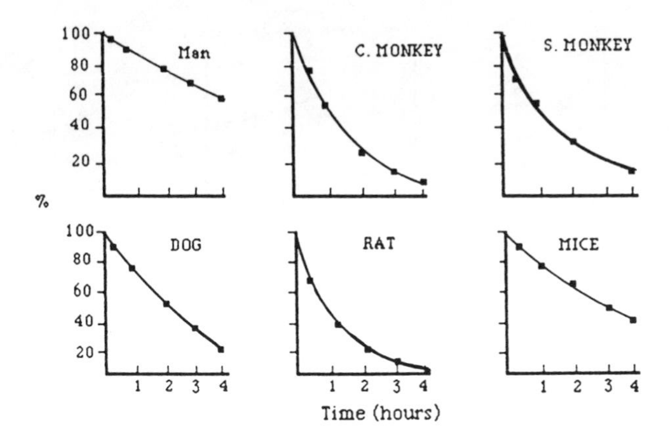

Figure 7. Stability of AZT-P-ddA in Plasma of Various Species at 37 C (4 ug/ml).

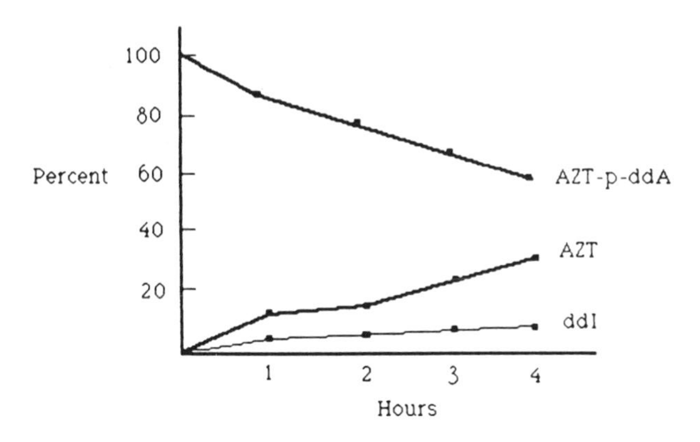

Figure 8. Disposition of AZT-P-ddA in Human Plasma at 37° (4 ug/ml).

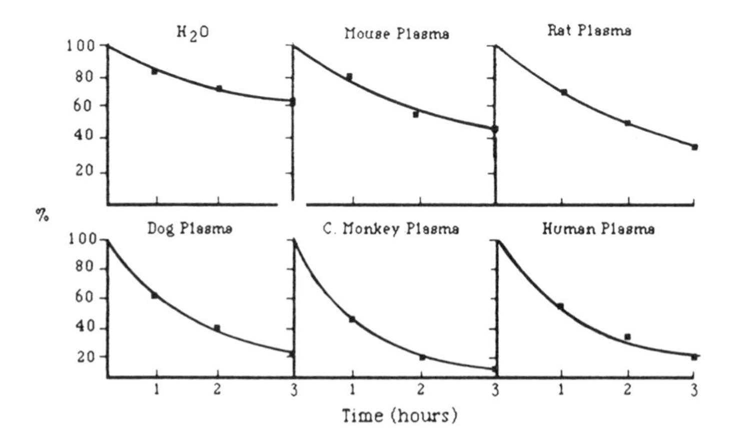

Figure 9. Stability of AZT-P(CyE)ddA at 37°C.

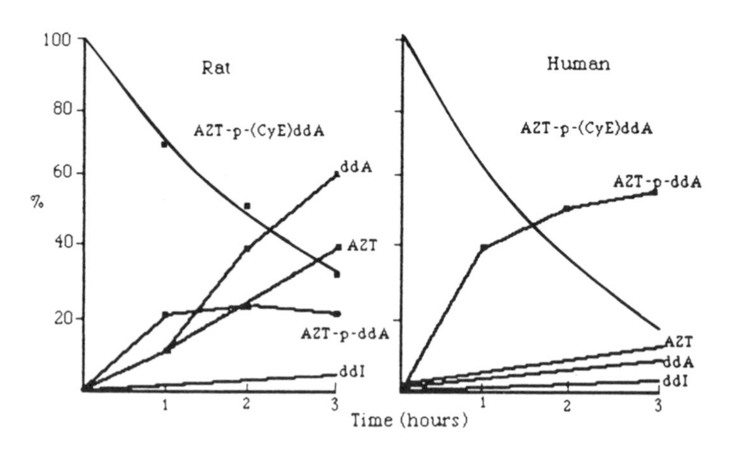

Figure 10. Stability of AZT-P(CyE)ddA in Human and Rat Plasma.

Table 2. Radioactivity in Rats Following I.V. Administration of AZT (5 mg/kg) AZT-P (CyE) ddA (11.5 mg/kg)

	Total Radioactivity (H^3) (% of dose)[a]			
Time	AZT		AZT-P (CyE)ddA	
	Plasma	Brain	Plasma	Brain
10 min	7.1	0.069	6.6	0.073
20 min	4.1	0.056	4.6	0.063
40 min	2.9	0.047	2.5	0.032
1 hr	1.4	0.030	1.4	0.028
2 hr	1.1	0.028	0.7	0.022
3 hr	0.5	0.024	0.6	0.026

a. Percent of dose was calculated by dividing radioactivity present in plasma or brain by total radioactivity administered

To assess in vivo disposition of AZT-P-ddA we synthesized radiolabelled substrate containing ^3H in a biologically stable position in the AZT molecule. This compound was diluted with unlabelled drug and administered intravenously (6mg/kg) into rats. Blood samples were obtained at 1, 2 and 3 hours after injection and analyzed for radioactive content. The results showed about 5.5% of the dose was present in serum after 1 hour. This value remained constant over the 3 hour sampling period. A similar study was carried out using radiolabelled AZT-P(CyE)-ddA and the results were compared to AZT (Table 2). The data shows that at each time period examined the concentration of both drugs present in plasma was not significantly different.

Additional studies are ongoing to determine whether *in vivo* advantages exceeding those observed *in vitro* may be associated with administration of the dimers. This possibility arises because first they do not have to contend with the fate of the individual components in terms of *in vivo* metabolism, and second, as intact nucleosides both components will reach the cell simultaneously.

Acknowledgment: The authors would like to thank Dr. C.C. Lin for his collaboration.

LITERATURE CITED

1. M. S. Gottlieb, R. Schroff, H. M. Schanker, J. D. Weisman, P. T. Fan, F. A. Wolf and A. Saxon, New Engl. J. Med. 305, 1425 (1981).

2. H. Masur, M. A. Michelis, J. B. Greene, I. Onorato, R. A. Vande Stouwe, R. S. Holzman, G. Wormser, L. Brettman, M. Lange, H. W. Murray and S. Cunningham-Rundles, New Engl. J. Med. 305, 1431 (1981).

3. F. P. Siegal, C. Lopez, G. S. Hammer, A. E. Brown, S. J. Kornfield, J. Gold, J. Hassell, S. Z. Hirschman, Parham, M. Siegal, S. Cunningham-Rundles and D. Armstrong , New Engl. J. Med. 305, 1439 (1981).

4. F. Barre-Sinoussi, J. C. Chermann, R. Rey, M. T. Nugeyre, S. Chamaret, J. Gruest, C. Dauguet, C. Axler-Blin, F. Brun-Vezinet, C. Rouzioux, W. Rozenbaum and L. Montagnier, Science 220, 868 (1983).

5. R. C. Gallo, S. Z. Salahuddin, M. Popovie, G. M. Shearer, M. Kaplan, B. F. Haynes, T. J. Palker, R. Redfield, J. Oleske, B. Safai, G. White, P. Foster and P. D. Markham, Science 224-500 (1984).

6. S. Broder and R. C. Gallo, New Engl. J. Med. 311. 1292 (1984).

7. M. Popovie, M. G. Sarngadharan, E. Read and R. C. Gallo, Science 224, 497 (1984).

8. F. Brun-Vezinet, C. Katlama, D. Roulot, L. Lenoble, M. Alizon, J. J. Madjar, M.A. Rey, P. M. Ginard, P. Yeni, F. Clavel, S. Gadelle, and M. Harzic. Lancet i:128 (1987).

9. F. Clavel, D. Guetard, F. Brun-Vezinet, S. Chamaret, M. A. Rey, M. O. Santos-Ferriera, A. G. Laurent, C. Dauget, C. Katlama, C. Rouzioux, D. Klatzman, J. L. Champalimaud, and L. Montagnier, Science 233:34 (1986).

10. E. DeClercq TIPS 8,39 (1987).

11. H. Mitsuga and S. Broder Nature 325, 773 (1987).

12. D. Baltimore Nature; 226:1209 (1970).

13. H.M Temin, S. Mizutani, Nature; 226:1211 (1970).

14. H. Mitsuya, R. Jarret, M. Matsukura, et al Proe Natl. Acad Sci, USA 84, 2033 (1987).

15. S. Broder, R. Yarchoan, J. M. Collins, H. C. Lane, P. D. Markham, R. W. Klecker, R. R. Redfield, H. Mitsuya, D. F. Hoth, E. Gelmann, J. E. Groopman, L. Resnick, R. C. Gallo, C. E. Myers and A. S. Fauci; Lancet ii, 627 (1985).

16. R. Yarchoan, R. W. Klecker, K. J. Weinhold, P. D. Markham, H. K. Lyerly, D. T. Durack, E. Gelmann, S. Nusinoff Lehrman, R. M. Blum, D. W. Barry, G. M. Shearer, M. A. Fischl, H. Mitsuya, R. C. Gallo, J. M. Collins, D. P. Bolognesi, C. E. Myers and S. Broder; Lancet i, 575 (1986).

17. R. Yarchoan, G. Berg, P. Brouwers, M. A. Fischl, A. R. Spitzer, A. Wichman, J. Grafman, R. V. Thomas, B. Safai, A. Brunetti, C. F. Perno, P. J. Schmidt, S. M. Larson, C. E. Myers and S. Broder; Lancet i, 132 (1987).

18. M. A. Wagar, M. J. Evans, K. F. Manly, et al. J. Cell Physiol. 121:402 (1984).

19. G. M. Tener, J. Amer. Chem. Suc. 83, 159 (1961).

20. D. D. Richman. Antimicrob. Ag. Chemoth. 31, 1879 (1987).

RECEIVED April 17, 1989

Chapter 12

Oligonucleotide Analogues as Potential Chemotherapeutic Agents

G. Zon

Applied Biosystems, Inc., 850 Lincoln Centre Drive, Foster City, CA 94404

In principle, relatively short oligonucleotides (ca. 15-20 bases in length) can specifically hybridize with DNA or RNA and thus be used for novel drug design strategies involving targeted interference of genetic expression at the level of transcription or translation. Conceivable chemotherapeutic applications predicated on sequence-specific hybridization ("antisense" inhibition) require oligonucleotide analogues that are resistant to in vivo degradation by enzymes such as nucleases. Nuclease-resistant analogues having modified internucleoside linkages (e.g., methylphosphonates or phosphorothioates) or modified nucleosides (e.g., 2'-0-methylribose or 1'-alpha-anomers) are now readily available by means of automated synthesis. Through collaborative investigations of various different backbone-modified oligonucleotides, we have comparatively studied hybridization (Tm), in vitro inhibition of gene expression (CAT and ras genes), and in vitro inhibition of HIV. Phosphorothioate oligomers at ca. 1 uM exhibited potent anti-HIV activity. The mechanism(s) for this anti-retroviral activity is (are) under further investigation.

Normal cell growth and replication require transcription of DNA and subsequent translation of mRNA to afford necessary enzymes, structural components, regulatory factors, and other types of proteins. These biochemical processes and protein products in aberrant or foreign form may lead to disease, as the end result of either an

0097–6156/89/0401–0170$06.00/0

inherited genetic defect, mutagenesis, or viral infections. A potentially general therapeutic strategy in such cases involves inhibition of aberrant or foreign ("target") gene expression by use of synthetic oligonucleotide analogues that have sequences complementary to specific sequences known to be present in the target DNA or RNA. Substrategies include targeting these single-stranded (ss) analogues at either double-stranded (ds) DNA, thus forming triplexes, or ss DNA or RNA, thus forming duplexes. Alternatively, ss or ds oligonucleotide analogues could be targeted at regulatory proteins that bind nucleic acids. Early studies (1,2) in this general area began more than two decades ago; however, only within the last few years has there been significantly increased attention and effort, perhaps due in part to recent advances in molecular biology, sequencing techniques, and automated synthesis of oligonucleotides. Moreover, there have been encouraging results in various antiviral model studies: Rous sarcoma virus (3), herpes simplex virus (HSV) type 1 (4), vesicular stomatitis virus (5), human immunodeficiency virus (HIV) (6-8), and type A influenza virus (9). A number of review articles (10-14) on such "antisense oligonucleotides" have appeared recently and can be consulted for much more information than can be given here. The present report is limited to an introduction to concepts in the design of anti-RNA oligomers, brief comments on their preparation and properties, and presentation of preliminary results obtained with oligonucleotide analogues containing phosphorothioate linkages, which are under development by the author through various collaborative investigations. This account will hopefully provide medicinal and carbohydrate chemists with an indication of the current potential, and problems to address, in the development of antisense oligonucleotides as a new class of antiviral agents.

Design of Anti-RNA Oligonucleotide Analogues

Important Design Factors. Therapeutic applications of oligonucleotides by means of sequence-specific inhibition of the function of RNA require consideration of the following factors. Practical methods for synthesis and purification of relatively large amounts of structurally modified oligonucleotides (analogues) that have adequate resistance to depolymerization by nucleases, adequate bioavailability, and are taken up by cells. The complementary target sequence must be accessible to the analogue and there must be an adequate rate and extent of association to form the resultant, perfectly base-paired duplex, which adequately inhibits the intended RNA-function. There should be minimal side-effects (e.g., that result from a finite amount of non-specific binding to polynucleotide sequences having partial homology, relative to the analogue-target duplex. The same is true for non-specific binding to other cellular constituents or enzymes. Formally analogous design criteria obtain for conventional (small molecule) antiviral agents; nevertheless, because of the comparatively unique nature of antisense oligonucleotides, it is worthwhile to consider these criteria in a bit more detail.

Structural Modifications. Oligonucleotides undergo nuclease-mediated depolymerization at rates which can be quite fast (half-life < 1h) in serum (15). Substitution of one of the non-bridging oxygens in the naturally occurring linkage (I) with either sulfur (II)(16), methyl (III) (11), amino (IV)(17) or alkoxy (V)(18) affords chemically stable, nuclease-resistant oligonucleotide analogues.

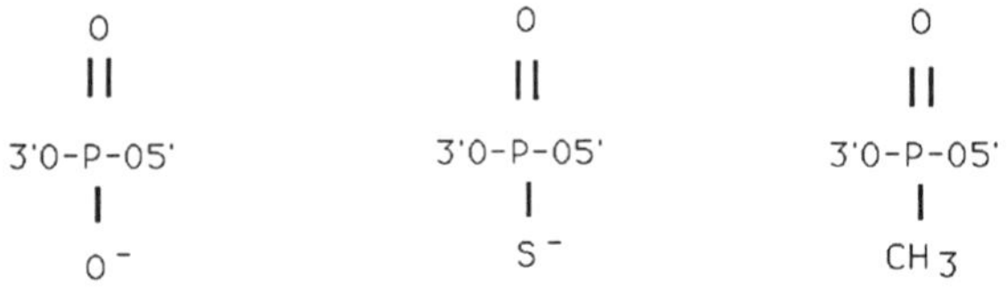

I	II	III
O	O	O
‖	‖	‖
3'O-P-O5'	3'O-P-O5'	3'O-P-O5'
|	|	|
O^-	S^-	CH_3

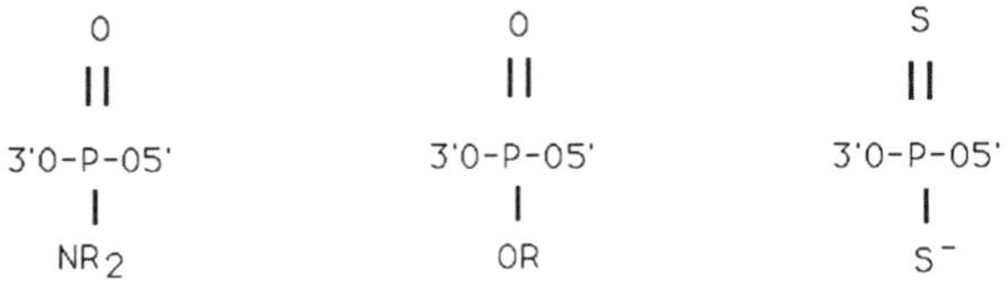

IV	V	VI
O	O	S
‖	‖	‖
3'O-P-O5'	3'O-P-O5'	3'O-P-O5'
|	|	|
NR_2	OR	S^-

Rather than modify every linkage, only several of these modifications at the 5' and 3' ends of an oligomer may provide adequate stability toward exonucleases (19), although endonucleolytic cleavage still occurs. This provides a means for controlling both the biological half-life of an oligonucleotide analogue and its partition coefficient. Additionally, in view of the chirality at phosphorus in II-V, which results in 2^n diastereomers (n=number of chiral linkages), use of only terminal modifications reduces the stereoisomeric heterogeneity of an oligonucleotide analogue. Another approach along this line is to use achiral linkages, such as VI (20). Achiral linkages without phosphorus have also been studied, namely, carboxymethyl (2), 3'O-$CH_2C(O)$-O5', carbamate (21,22), 3'O-C(O)-NH5', and sulfide (23), 3'C-CH_2CH_2S-C5', moieties.

Among the modified nucleosides that might provide adequate resistance to nucleases, oligomers containing alpha-2'-deoxynucleoside anomers have attracted considerably more attention (24,25) than those with other types of stereoisomerism (14) or with 2'-O-methylribonucleosides (26). Oligonucleotides with base modifications that impart resistance to nucleases apparently have not been reported.

Bioavailability and Uptake by Cells. Miller and Ts'o have reported (11) that methylphosphonate analogues (III) prevent expression of herpetic lesions when applied in the form of a cream to the HSV-infected ear of a mouse. A tritium-labeled methylphosphonate analogue was found (11) to be distributed to all organs and tissues, with the exception of the brain, when injected into the tail vein of a mouse. By contrast, in a similar study with carbon-labeled methylphosphonate analogues, Maguire and Han (27) reported that, on a wet weight basis, peak radioactivities were in the order: brain > liver > kidney > lung > heart > muscle. The occurrence times for these peaks were in the order: muscle > brain = heart = lung > liver > kidney. Indirect evidence for adequate bioavailability of oligonucleotides with unmodified linkages, and their uptake by cells, is provided by reports of in vivo activity in mice against HSV-1 (28), tick-borne encephalitis virus (29), and c-myc (30). Uptake by cells can be studied with radiolabeled oligonucleotides, by either counts alone or autoradiography to visualize the distribution. Alternatively, one can use fluorescently labeled oligomers, with either fluorescence-activated cell sorting (FACS) or fluorescence microscopy. Methylphosphonate analogues are thought to be taken up by passive diffusion (11) whereas uptake of negatively charged, unmodified oligonucleotides and phosphorothioate (II) analogues may involve a more complex process. Our preliminary FACS measurements (Egan, W., Food and Drug Administration, personal communication, 1988) with the latter analogues indicate a very rapid (<10 min) increase in cell-associated fluorescence followed by further increase very slowly or an apparent plateau. At this time it is unknown whether this represents uptake alone or involves an initial fast association with positively charged groups on the cell surface. A potentially major discovery in this area involves a putative oligonucleotide receptor (Stein, C., National Institutes of Health, personal communication, 1988). Also noteworthy are several promising schemes for obtaining facilitated uptake: conjugation of oligonucleotides to poly(L-lysine) (31) or steroids (32) and packaging of oligonucleotides in liposomes (29,33), Sendai virus envelope (29), or erythrocyte "ghosts" (29).

Specific Binding to Target and Inhibition of Function. In 1978 Zamecnik and coworkers (3) coined the word "hybridon" to refer to competitive hybridization of synthetic oligonucleotides to target sequences in order to block either (1.) RNA→DNA polymerization (reverse transcription), (2.) DNA integration, (3.) DNA→RNA transcription, (4.) translation, or (5.) ribosomal association. A splice-junction can also be targeted (4). Calculations of RNA tertiary structure to determine accessible ss target sequences have been carried out (34); however, these are of questionable value, and most reports to date have tested oligomer sequences complementary to regions either near or at the start of translation, which are assumed to be generally accessible. An important development has been recognition that formation of oligodeoxyribonucleotide-RNA heteroduplexes may lead to scission of the ribo strand by RNase H (35), which apparently is present in all (or most) cells. This allows for catalytic turnover of RNA as opposed to 1:1 stoichiometric blockage. Significantly, phosphorothioate but not alpha-anomeric oligonucleotide analogues can function efficiently in this manner (36).

Statistically, a sequence of ca. 17 nucleotides should be found only once in the human genome, and shorter sequences (ca. 12-mers) can, in principle, be of use at the level of mRNA. From DNA probe technology, it is reasonable to infer that one base pair mismatch in an RNA target-oligonucleotide analogue duplex causes sufficient destabilization so as to favor a high degree of binding specificity, in this range of lengths. But this is true only under thermodynamic control (at equilibrium) and so-called stringent conditions (relatively high temperature, low salt concentration, and low concentration of oligomer). In contrast, uptake of an oligonucleotide analogue by a live cell in vivo involves non-equilibrium conditions, with regard to formation of oligomer-RNA (or DNA) duplexes, and non-adjustable non-stringent conditions. This indicates that the desired degree of specificity of duplex formation may be offset by kinetically controlled phenomena and the rate at which the "system" approaches a state of pseudo-equilibrium. It may therefore be possible to maximize specificity, at least in part, by adopting a strategy in which "shorter is better". Yet to be determined is whether the synergy of contiguous antisense oligomers evidenced in cell-free studies (37) extends to applications in vivo.

Schemes for either decreasing or preventing disassociation of an antisense oligomer from its target, or displacement by polynucleotide/protein, include the use of pendant alkylating groups, intercalators, and photoactivatable groups (10-14).

Preparation and Physical Characteristics of Oligonucleotide Analogues

Synthesis and Purification. Oligodeoxyribonucleotides and oligoribonucleotides are readily prepared by automated versions of the solid-phase phosphoramidite method developed by Caruthers and coworkers (38) (Figure 1). In each cycle, coupling affords a phosphite linkage which can be either oxidized with I_2/H_2O/base to a phosphotriester precursor of an unmodified linkage, or sulfurized with S_8/CS_2/pyridine, or converted with I_2/R_2NH to a phosphoramidate (IV), or stable phosphotriester (V) (39). Obvious extension to methylphosphonamidites (40) affords methylphosphonate linkages. It is thus possible to introduce into an oligonucleotide analogue one or more of the same or different modifications in any order. The ability to assemble oligomers with combinations of charged and neutral linkages proved to be useful, as our initial work with methylphosphonate and isopropylphosphotriester analogues indicated low solubility in water, compared to charged oligomers, leading to problems with handling and biological testing, which were addressed by synthesis of n-mer analogues having alternating linkages: (I-III)n/2 and (I-V)n/2. Another approach to adjusting solubility is the use of aminoalkyl substituents attached to either phosphorus (41) or base moieties. Nielsen and Caruthers (42) have described regioselective Arbuzov-type reactions of internucleoside 2-cyano-1,1-dimethylethylphosphites as a novel and versatile route to mixed-linkage analogues. Such molecules can also be constructed by appropriate alternation of phosphoramidite and H-phosphonate chemistry cycles (Zon, G., unpublished data), the latter of which features coupling by activation of the H-phosphonate monomer with adamantane carbonyl chloride to give a mixed anhydride, and capping by similar reaction with isopropyl phosphite (43) (Figure 2). The resultant internucleoside H-phosphonate linkage(s) can be either oxidized to an unmodified linkage or converted into modified linkages such as II, IV, V (17-44). Stec and coworkers (45) have reported the use of a bis-amidite reagent as an alternative entry to either phosphoramidite- or H-phosphonate-like chemistry during each cycle of chain extension.

Polyacrylamide gel electrophoresis has been traditionally used to purify synthetic oligonucleotides

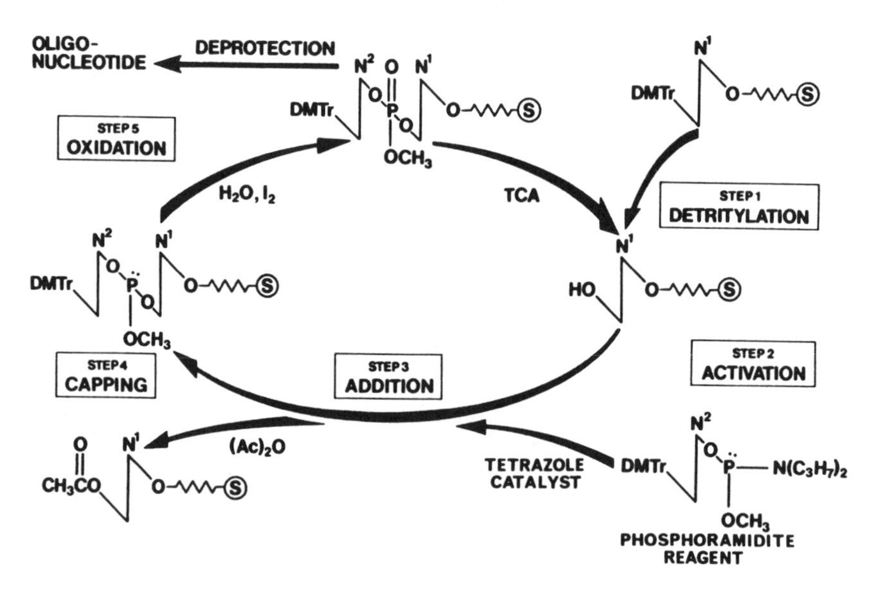

Figure 1. Schematic representation of the phosphoramidite method for oligonucleotide synthesis on a solid support. For definitions and details, see ref. 38.

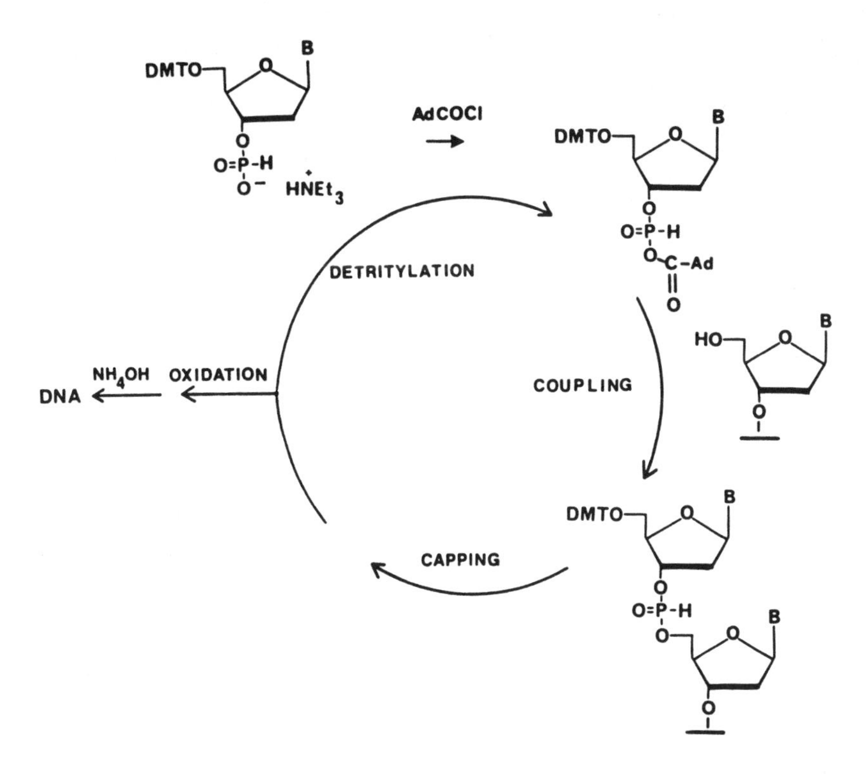

Figure 2. Schematic representation of the hydrogen phosphonate method for oligonucleotide synthesis on a solid support. For definitions and details, see ref. 43.

in small amounts (< 1 mg), while ion-exchange HPLC and reversed-phase (paired-ion) HPLC have become more prevalent methods of purification because they can be scaled-up and automated. Recent reviews (46) of oligonucleotide purification can be consulted for details. An increasing gradient of acetonitrile in triethylammonium acetate is frequently used for reversed-phase HPLC of oligonucleotides because the eluents are relatively easy to remove from the final product. High-resolution proton NMR can be employed to check for residual eluents, while phosphorus NMR is useful for analysis of the chemical homogeneity of the oligomer backbone when appropriate.

Stereochemistry and Strength of Binding. A plot of UV absorbance at 260 nm (A_{260}) vs. temperature generally gives an "S"-shaped curve (melt profile) that has an inflection point that indicates the temperature (Tm) at which a duplex is half dissociated, at the specified concentration of oligomers, added salt (cation) and pH. A plot of Tm vs. log[cation activity] provides thermodynamic constants for duplex formation, which can be interpreted in terms of the strength of binding. The influence on Tm of converting an unmodified phosphodiester linkage (I) into a chiral linkage, 3'O-P(O)(X)-O5', should depend on the electronic nature of X, its steric "bulk", and the absolute stereochemistry at phosphorus. An electronically neutral (e.g., X=methyl) or positively charged substituent (e.g., X=aminoalkyl) should decrease the dependence of Tm on the cation concentration, but apparently it is not known whether this is beneficial under physiological salt conditions. With regard to absolute stereochemistry at phosphorus (R or S configuration), inspection of molecular models of duplexes suggests two different orientations of substituent X, relative to the helical axis, each having different local interactions with the flanking residues and associated water molecules. We have utilized a series of backbone modified, self-complementary octadeoxyribonucleotides (e.g., R or S $GGAA_XTTCC$) to establish the absolute configurations at phosphorus, by chemical and NMR methods, and measure the change in Tm (ΔTm) associated with the "outward" and "inward" orientations of X, relative to the unmodified duplex. The latter orientation for X=sulfur (47), methyl (48), or alkoxy (49) depresses the Tm, which has been ascribed to repulsive steric interactions between X and the 3' proton.

The influence on Tm of multiple methylphosphonate modifications, and the influence on Tm of the nature of the flanking bases, has been investigated (Wilson, W. D., Georgia State University, personal communication,1988) in the self-complementary tetradecadeoxyribonucleotide

T(AATT)$_3$A, which when unmodified has Tm = 35.8°C in 0.16 M NaCl, as noted in Table I. Separated modifications in this series, relative to clustered ones, have a somewhat less disruptive effect on helical stability. Incorporation of twelve contiguous modifications lowered the Tm by only 12.6°C.

Table I. Location of Methylphosphonate Linkages (X), and Corresponding Tm Values at 0.16 M NaCl

Tm (°C)	T		A		A		T		T		A		A		T		T		A		A		T		T		A
35.8	unmodified																										
31.7				X								X								X							
30.5						X							X								X						
30.4								X								X								X			
32.2										X								X								X	
33.5														X		X		X									
20.9				X		X		X		X		X		X		X		X		X		X		X		X	

We have obtained other data of this sort for additional sequences and for phosphorothioate linkages (II), which likewise cause a roughly 10°C drop in Tm when arranged contiguously in 10- to 30-mers. Both methylphosphonate and phosphorothioate analogues show Tm curves that are displaced but not significantly broadened, relative to the unmodified duplex. The latter feature might have been unexpected, in view of the presence of 2^n diastereomers, depending on one's theoretical analysis and assumptions.

Anti-HIV Activity of Phosphorothioate Analogues

Preliminary Studies. Among the goals of our initial studies were comparison of the efficacy of various types of analogues and the use of quantitative assays. At that time the neutral methylphosphonates and charged, unmodified oligomers were known (3,4) to be antisense inhibitors of protein synthesis, and it was of interest to compare them with partially charged methylphosphonates and phosphotriesters, as well as charged phosphorothioates. We chose a relatively simple eukaryotic tissue culture (CV-1 cells) transfection assay of plasmid containing the gene for expression of the bacterial enzyme chloramphenicol acetyltransferase (CAT) coupled to viral (SV 40) regulatory sequences as a model in vitro system for gauging the relative effectiveness of these analogues in the regulation of expression of a specific protein (50). The target was a 21-base sequence that began with AUG, which codes for the initial methionine. Transfection in the presence of 30 μM oligomer led to the following order of oligomer inhibitory effect: phosphorothioate > methylphosphonate > alternating methylphosphonate and unmodified > alternating ethylphosphotriester and unmodified > unmodified = alternating isopropylphosphotriester and unmodified, with the first case being 84% and the last two cases being 35% and 0%, respectively.

A cell-free translation assay in rabbit reticulocyte lysate, which had been developed for evaluating methylphosphonate analogues as inhibitors of ras-p21, was made available to us for comparative studies with phosphorothioates (Chang, E. H., Uniformed Services University of Health Science, personal communication, 1988). An 11-mer methylphosphonate complementary to the 5' end of the mRNA caused ~45% and ~95% inhibition of p21 translation at 50 μM and 200 μM, respectively. Oligomers having unmodified linkages or alternating unmodified and methylphosphonate linkages were less effective, whereas the phosphorothioate analogue caused ~60% and ~100% inhibition at 12.5 μM and 50 μM, respectively. The specificity of the phosphorothioates in this system is still under investigation.

Anti-HIV Activity In Vitro. Based on the presumed importance of the art/trs (rev) encoded protein in the regulation of HIV, a 14-mer sequence and various presumed irrelevant (control) sequences were studied in a cytopathic effect inhibition assay (7). None of the tested methylphosphonate and unmodified sequences showed statistically significant activity, whereas to our surprise all of the phosphorothioates (except very short ones, 5-mers) exhibited some level of activity. A 28-mer polymer of dC was the most potent phosphorothioate tested, showing complete inhibition at 0.5 - 1.0 μM. Subsequent experiments with this material at 1 μM employed a Southern blot analysis to demonstrate complete inhibition of de novo viral DNA synthesis. Recent unpublished results suggest that phosphorothioate oligomers can inhibit RNA → DNA transcription by binding to viral reverse transcriptase (RT) (Cohen, J. S. National Institutes of Health, Private communication, 1988). In addition to this presumptive mechanism for *sequence-nonspecific*, RT-directed antiviral activity of phosphorothioate oligomers in non-infected cells, a *sequence-specific* mechanism has been found more recently in an assay for inhibition of viral expression of HIV in chronically infected T-cells (Matsukura, M., National Institutes of Health, private cummunication, 1988). The active phosphorothioate sequence is the extended (28-mer) version of the aforementioned anti-art/trs (rev) oligomer. The results of Northern blot analysis suggest that there are marked effects on RNA; however, the details have yet to be established. In any event, at this time the phosphorothioate analogues are especially interesting as potential anti-retroviral agents in that they that may have two, independent mechanisms of action which operate at different points in the virus life-cycle.

Literature Cited

1. Belikova, A. M.; Zarytova, V. F., Grineva, N. I. Tetrahedron Lett. 1967, 3557.

2. Halford, M. H.; Jones, A. S. Nature 1968, 217, 638.

3. Zamecnik, P. C.; Stephenson, M. L. Proc Natl. Acad. Sci. USA 1978, 75, 280.

4. Smith, C. C.; Aurelian, L.; Reddy, M. P.; Miller, P. S.; Ts'o, P. O. P. Proc. Natl. Acad. Sci. USA 1986, 83, 2787.

5. Agris, C. H., Blake, K. R.; Miller, P. S. Reddy, M. P.; Ts'o, P.O.P. Biochemistry 1986, 25, 6268.

6. Zamecnik, P. C.; Goodchild, J.; Taguchi Y., Sarin, P. S. Proc. Natl. Acad. Sci USA 1986, 83 4143.

7. Matsukura, M.; Shinozuka, K.; Zon, G.; Mitsuya, H.; Reitz, M.; Cohen, J. S.; Broder, S. Proc. Natl. Acad. Sci. USA 1987, 84, 7706.

8. Goodchild, J.; Agrawal, S.; Civeira, M. P.; Sarin, P. S.; Sun, D.; Zamecnik, P. C. Proc. Natl. Acad. Sci., USA 1988, 85, 5507.

9. Zerial, A.; Thuong, N. T.; Helene, C. Nucleic Acids Res. 1987, 15, 9909.

10. Knorre, D. G.; Vlassov V. V.; Zarytova, V. F. Karpova, G. G. Adv. Enz. Reg. 1985, 24, 277.

11. Miller, P. S.; Ts'o, P. O. P. Anti-Cancer Drug Design 1987, 2, 117.

12. Paoletti, C. Anti-Cancer Drug Design 1988, 2, 325.

13. Stein, C. A.; Cohen, J. S. Cancer Res. 1988, 48, 2659.

14. Zon, G. Pharm Res. 1988, 5, 539.

15. Wickstrom, E. J. J. Biochem. Biophys. Methods 1986, 13, 97.

16. Stec, W. J.; Zon, G.; Egan. W.; Stec, B. J. Am. Chem. Soc. 1984, 106, 6077.

17. Froehler, B. C. Tetrahedron Lett. 1986, 27, 5575.

18. Stec, W. J.; Zon, G.; Gallo, K. A.; Byrd, R. A.; Uznanski, B.; Guga, P. Tetrahedron Lett. 1985, 26, 2191.

19. Agrawal. S.; Goodchild, J. Tetrahedron Lett. 1987, 28, 3539.

20. Brill, W. K.-D.; Caruthers, M. H. Abstracts Intl. Workshop on Perspectives in Therapeutic and Diagnostic Applications of Oligonucleotide Derivatives, 1988.

21. Coull, J. M.; Carlson, D. V.; Weith, H. L. Tetrahedron Lett. 1987, 28, 745.

22. Stirchak, E. P.; Summerton, J. E.; Weller, D. D. J. Org. Chem. 1987, 52, 4202.

23. Kawai, S. H.; Just, G.; Chin. J. Abstract Org. Chem. Third Chem. Congress of N. America, 1988, No. 318.

24. Sun, J.-S.; Asseline, U.; Rouzaud, D.; Montenay-Garestier, T.; Thoung, N. T.; Helene, C. Nucleic Acid Res. 1987, 15, 6149.

25. Gautier, C.; Morvan, F.; Rayner, B; Huynh-Dinh, T.; Igolen, J.; Imbach, J.-L.; Paoletti, C. Paoletti, J. Nucleic Acids Res. 1987, 15, 6625.

26. Inoue, H. Hayase, Y.; Iwai, S,; Ohtsuka, E. FEBS Lett. 1987, 215, 327.

27. Maguire, D. J.; Han, H. Proc. Australian Biochem Soc., 1988, P26.

28. U. S. Patent No. 4,689,320, 1987.

29. Vlassov, V. V. Abtsracts Intl. Workshop on Perspectives in Therapeutic and Diagnostic Applications of Oligonucleotide Derivatives, 1988.

30. Japanese Patent No. 61-122215, 1984.

31. Lemaitre, M.; Bayard, B.; Lebleu, B.. Proc. Natl. Acad. Sci. USA 1987, 84, 648.

32. Ivanova, E. M.; Chasovskikh, M. I.; Zarytova, V. F. Abstract Intl. Workshop on Perspectives in Therapeutic and Diagnostic Applications of Oligonucleotide Derivatives, 1988.

33. Loke, S. L.; Stein, C.; Avigan, M.; Cohen, J.; Neckers, L. M. Clin. Res. 1988, 36, 443A.

34. Wickstrom, E.; Simonet, W. S.; Medlock, K.; Ruiz-Robles, I. Biophys. J. 1986, 49, 15.

35. Walder, R. Y.; Walder, J. A. Proc. Natl. Acad. Sci. USA 1988, 85, 5011.

36. Cazenave, C.; Cohen, J. S.; Helene, C.; Loreau, N.; Stein, C. A.; Thoung, N. T.; Toulme, J. J. Abstracts Intl. Workshop on Perspectives in Therapeutic and Diagnostic Application of Oligoncleotide Derivatives, 1988.

37. Maher, L. J.,III; Dolnick, B. J. Arch. Biochem. Biophys. 1987, 253, 214.

38. Caruthers, M. H. Science 1985, 230, 281.

39. Koziolkiewicz, M. Uznanski, B.; Stec, W. J.; Zon, G. Chem. Scr. 1986, 26, 251.

40. Applied Biosystems DNA Synthesizer Model 380 User Bulletin, Issue No. 43, October 1, 1987.

41. Letsinger, R. L.; Singman, C. N.; Histand, G.; Salunkhe, M. J. Am. Chem Soc. 1988,. 110, 4470.

42. Nielsen, J.; Caruthers, M. H. J. Am. Chem. Soc. 1988, 110, 6275.

43. Andrus, A.; Efcavitch, J. W.; McBride, L. J.; Giusti, B. Tetrahedron Lett. 1988, 29, 861.

44. Froehler, B.; Ng, P.; Matteucci, M. Nucleic Acids Res. 1988, 16, 4831.

45. Uznanski, B.; Wilk, A.; Stec, W. J. Tetrahedron Lett. 1987, 28, 3401.

46. Zon, G.; Thompson, J. A. BioChromatography 1986, 1, 22.

47. LaPlanche, L. A.; James, T. L.; Powell. C.; Wilson, W. D.; Uznanski, B.; Stec, W. J.; Summers, M. F.; Zon, G. Nucleic Acids Res. 1986 14, 9081.

48. Bower, M.; Summers, M. F.; Powell, C.; Shinozuka, K.; Regan, J. B.; Zon, G.; Wilson, W. D. Nucleic Acids Res. 1987, 15, 4915.

49. Lawrence, D. P.; Wenqiao, C.; Zon, G.; Stec, W. J.; Uznanski, B.; Broido, M. S. J. Biomol. Struct. Dyn. 1987, 4, 757.

50. Marcus-Sekura, C. J.; Woerner, A. M.; Shinozuka, K.; Zon, G.; Quinnan, G. V. Jr. Nucleic Acid Res. 1987, 15, 5749.

RECEIVED January 4, 1989

Author Index

Affiliation Index

Subject Index

A

R

S

T

Production: Colleen P. Stamm
Indexing: Deborah H. Steiner
Acquisition: Cheryl Shanks

Elements typeset by Hot Type Ltd., Washington, DC
Printed and bound by Maple Press, York, PA

Other ACS Books

Chemical Structure Software for Personal Computers
Edited by Daniel E. Meyer, Wendy A. Warr, and Richard A. Love
ACS Professional Reference Book; 107 pp;
clothbound, ISBN 0–8412–1538–3; paperback, ISBN 0–8412–1539–1

Personal Computers for Scientists: A Byte at a Time
By Glenn I. Ouchi
276 pp; clothbound, ISBN 0–8412–1000–4; paperback, ISBN 0–8412–1001–2

Biotechnology and Materials Science: Chemistry for the Future
Edited by Mary L. Good
160 pp; clothbound, ISBN 0–8412–1472–7; paperback, ISBN 0–8412–1473–5

Polymeric Materials: Chemistry for the Future
By Joseph Alper and Gordon L. Nelson
110 pp; clothbound, ISBN 0–8412–1622–3; paperback, ISBN 0–8412–1613–4

The Language of Biotechnology: A Dictionary of Terms
By John M. Walker and Michael Cox
ACS Professional Reference Book; 256 pp;
clothbound, ISBN 0–8412–1489–1; paperback, ISBN 0–8412–1490–5

Cancer: The Outlaw Cell, Second Edition
Edited by Richard E. LaFond
274 pp; clothbound, ISBN 0–8412–1419–0; paperback, ISBN 0–8412–1420–4

Practical Statistics for the Physical Sciences
By Larry L. Havlicek
ACS Professional Reference Book; 198 pp; clothbound; ISBN 0–8412–1453–0

The Basics of Technical Communicating
By B. Edward Cain
ACS Professional Reference Book; 198 pp;
clothbound, ISBN 0–8412–1451–4; paperback, ISBN 0–8412–1452–2

The ACS Style Guide: A Manual for Authors and Editors
Edited by Janet S. Dodd
264 pp; clothbound, ISBN 0–8412–0917–0; paperback, ISBN 0–8412–0943–X

Chemistry and Crime: From Sherlock Holmes to Today's Courtroom
Edited by Samuel M. Gerber
135 pp; clothbound, ISBN 0–8412–0784–4; paperback, ISBN 0–8412–0785–2
